T0135779

Augsburger Schriften zur Mathematik, Physik und Informatik
Band 15

herausgegeben von:
Professor Dr. F. Pukelsheim
Professor Dr. W. Reif
Professor Dr. D. Vollhardt

Bibliografische Information der Deutschen Nationalbibliothek

Die Deutsche Nationalbibliothek verzeichnet diese Publikation in der
Deutschen Nationalbibliografie; detaillierte bibliografische Daten sind
im Internet über http://dnb.d-nb.de abrufbar.

ISBN 978-3-8325-2815-7
ISSN 1611-4256

Logos Verlag Berlin GmbH
Comeniushof, Gubener Str. 47,
10243 Berlin
Tel.: +49 030 42 85 10 90
Fax: +49 030 42 85 10 92
INTERNET: http://www.logos-verlag.de

Adaptive Finite Elements in the Discretization of Parabolic Problems

Dissertation zur Erlangung des Doktorgrades der
Mathematisch-Naturwissenschaftlichen Fakultät der
Universität Augsburg

vorgelegt von Christian A. Möller

November 2010

Erster Gutachter: Prof. Dr. K. G. Siebert, Duisburg

Zweiter Gutachter: Prof. Dr. M. A. Peter, Augsburg

Mündliche Prüfung: 23. Dezember 2010

Dank

An dieser Stelle möchte ich mich sehr herzlich bei allen bedanken, die zum Gelingen dieser Arbeit beigetragen haben und mich im Laufe der letzten Jahre unterstützt haben. An erster Stelle gebührt mein Dank meinem Doktorvater Kunibert G. Siebert, der mich schon seit den frühen Semestern meines Studiums hervorragend betreut und unterstützt hat. Er hat mich in den spannenden Themenkreis der adaptiven Finite Elemente Methoden eingeführt und war — in den letzten beiden Jahren auch aus der Ferne — bei Fragen stets zur Stelle. Außerdem bedanke ich mich herzlich bei Malte Peter für die spannende und überaus angenehme Zusammenarbeit sowie die freundliche und kompetente Unterstützung bei allen Fragen. Ebenso danke ich allen Kollegen, insbesondere Christian Kreuzer und Johannes Neher, die immer Zeit für Diskussionen hatten und die Teile dieser Arbeit Korrektur gelesen und wertvolle Verbesserungsvorschläge eingebracht haben.

Außerdem möchte ich mich bei meinem Studiengang TopMath und dem Elitenetzwerk Bayern bedanken, die Besuche von Konferenzen und Vorträgen unterstützt und mir auch jenseits der Mathematik die Teilnahme an vielen interessanten Softskill-Seminaren ermöglicht haben. Der Universität Bayern e.V. und der Studienstiftung des deutschen Volkes danke ich vielmals für die großzügige und unkomplizierte finanzielle Unterstützung.

Zu guter Letzt ein herzliches Dankeschön an meine Familie und Freunde, durch die ich die Freiheiten der letzten Jahre erst richtig genießen konnte, die es aber auch in schwierigen Zeiten immer geschafft haben, mich wieder zu motivieren. Michi, wenn ich Dir hier für alles einzeln danken würde, enthielten die nächsten 240 Seiten zu wenig Mathematik... Danke!

Contents

Chapter 1

Introduction

Adaptive finite element methods have become an essential numerical tool for the solution of partial differential equations over the last decades. Particularly interesting in this field is the featured adaptivity, which allows an adaptive finite element method (AFEM) to solve a given problem efficiently. The basis for such adaptive methods are so-called a posteriori error estimators, which allow to gain accuracy information of a provided approximate solution of a given partial differential equation (PDE). Most importantly, in this process the exact solution — which is usually unknown — is *not* needed, but rather, only the discrete approximation and given data of the PDE are considered. Information received from such a posteriori error estimators is used twofold: Firstly, knowledge about the local distribution of the discretization error allows for deriving methods suited for effectively decreasing the global error and secondly, information about the overall discretization error is employed to stop an adaptive procedure once the desired accuracy is reached.

A particularly interesting question in this context regards termination and convergence: Can an AFEM reach any given tolerance within finitely many iterations? For linear elliptic problems, this has recently been answered positively by Morin, Siebert, and Veeser, cf. [36]. Building on this, the present thesis addresses the still unanswered question of termination and convergence for linear parabolic problems of the from

$$\begin{aligned}
\partial_t u(x,t) + \mathcal{L}u(x,t) &= f(x,t) && \text{in } \Omega \times (0,T), \\
u(x,t) &= 0 && \text{on } \partial\Omega \times (0,T), \\
u(\cdot,0) &= u_0(x) && \text{in } \Omega,
\end{aligned} \tag{1.1}$$

where Ω is a spatial domain, $(0,T)$ is a time interval, and $\mathcal{L}$ denotes an elliptic differential operator. In the simplest form of $\mathcal{L} = -\Delta$, equation (1.1) becomes the well-known heat equation, which is used for modeling various kinds of physical diffusion processes.

In order to solve (1.1) by an adaptive finite element method, a discretization of the problem has to be provided. In this work, we consider an implicit Euler discretization with respect to time, which splits the time dependent problem (1.1) into a sequence of stationary elliptic problems associated with discrete time nodes $t_0, \ldots t_N$ with $t_n \in [0,T]$ for $n = 0, \ldots, N$. We then employ a time stepping to solve those problems one at a time using an AFEM for stationary problems. However, since we are interested in a fully adaptive method, we also wish to choose the discrete time nodes $t_0, \ldots, t_N$ adaptively. This means that having finished the n-th time step, the position of the next time node t_{n+1} is determined by adaptively choosing a time step size τ_{n+1} and setting $t_{n+1} := t_n + \tau_{n+1}$. Note that since the time step sizes are chosen adaptively, it is not clear whether the desired final time T will be reached, in particular we may have $N = \infty$ and premature convergence of the sum of time step sizes $\sum_{n=1}^{\infty} \tau_n = T' < T$ may occur. The process of adaptively choosing the new time step size τ_{n+1} involves a local time error indicator and τ_{n+1} is chosen such that the anticipated time discretization error on the time interval (t_n, t_{n+1}) is sufficiently small. However, having computed the solution of the stationary problem for time step $n + 1$, it might turn out that the time step size τ_{n+1} was chosen too big to provide appropriate error control. In this case, τ_{n+1} has to be decreased and t_{n+1} approaches t_n. This change of the time node t_{n+1} in turn changes the associated stationary problem. With this in mind, we realize that it is not clear at all that even a single time step terminates: From the elliptic convergence theory mentioned above, we are only able to guarantee that in case of a *fixed time node*, the associated elliptic problem can be solved for any desired accuracy within finitely many iterations.

The first step towards termination and convergence for an adaptive finite element method for time dependent problems was taken by Chen and Feng in [14], where they showed that eventually, each single time step reaches any demanded tolerance. This was accomplished by showing that for each time step n, there exists a minimal time step size τ_n^* which guarantees that a prescribed tolerance for the time discretization error is reached. Having at hand such a minimal time step size, we see that the associated elliptic problem will be fixed eventually. In this situation, elliptic convergence theory may be applied to guarantee that the considered time step eventually terminates while reaching its prescribed tolerance. However, since the minimal time step size τ_n^* is dedicated to one particular time step and the individual τ_n^* are unrelated for different time steps, premature convergence $\sum_{n=1}^{\infty} \tau_n^* = T' < T$ may still appear and it is not clear whether or not the final time T is reached.

To circumvent this problem, we slightly generalize the concepts of [27] to a more general problem class and introduce a special kind of error control, which allows for employing a *globally* smallest time step size τ_* while still being able to reach any given tolerance. The minimal time step size τ_* only depends on the desired tolerance and data of the problem and since it is valid globally (i. e., in each time step), it is clear that the final time T is reached after at most T/τ_* time steps. We implement this error control in an adaptive space time finite element method (ASTFEM) which reaches any prescribed tolerance within finitely many iterations.

Based on the algorithm ASTFEM, we also consider the real life problem of concrete carbonation. More precisely, we simulate the diffusion of atmospheric CO_2 into a concrete structure which evokes a chemical reaction lowering the pH. As a model underlying the simulation, we use the homogenization approach of [49]. This is of particular interest since the lowered pH facilitates the corrosion of steel reinforcements embedded in the concrete. The simulation tremendously profits from the employed adaptivity since the simulated quantities exhibit locally very inhomogeneous structures.

1.1   Outline of the thesis

This work starts from the analytic fundamentals in Chapter 2 where we introduce the class of linear parabolic partial differential equations considered throughout this thesis by adding details to (1.1). We derive a weak formulation and establish existence of a unique solution.

This weak formulation is the basis of the discretization presented in Chapter 3. In this chapter, we provide basics of finite element spaces and the underlying triangulations. We then present a full space time discretization of the considered problem. In this process, the time dependent problem is converted into a sequence of stationary problems as indicated above. Motivated by this, the whole Section 3.3 is devoted to elliptic problems and ranges from existence and uniqueness results to error estimates and finally to the prototype of an adaptive algorithm for stationary problems. After this digression into elliptic theory, we turn back to the time dependent problem and present a posteriori error estimates.

In Chapter 4 we then build on the derived a posteriori estimates to construct an adaptive space time finite element method (ASTFEM). In the process, we introduce a new kind of error control which is based on a uniform energy estimate and allows for employing a global minimal time step size bounded away from zero. This enables us to guarantee that ASTFEM terminates while providing appropriate error control. Chapter 4 covers these topics employing abstract modules as components of ASTFEM before suggesting specific realizations of these modules.

Chapter 5 focuses on numerical aspects of ASTFEM. In particular, multi-coarsening is introduced as a new coarsening strategy. Furthermore, we compare ASTFEM to a standard adaptive algorithm for parabolic problems and present numerical results, which are based on an implementation using multi-coarsening.

The final Chapter 6 is devoted to applying ASTFEM to the real life problem of carbonation of concrete. After presenting a homogenized model of concrete carbonation which was originally derived in [49], we consider two types of discretizations. Even though these problems exceed the theoretical framework derived for ASTFEM in Chapters 3 and 4, we adapt ASTFEM heuristically and present simulations of concrete carbonation. We compare the obtained results to experimental data and also consider differences of the two employed discretizations.

First of all, however, we fix some notation used throughout this work and review some basic properties of the employed function spaces in the next section.

1.2 Notation and function spaces

In this work we choose $\mathbb{R}$ to denote the set of real numbers, $\mathbb{N}$ as the set of natural numbers as well as $\mathbb{N}_0 := \mathbb{N} \cup \{0\}$. For any set $A \subset \mathbb{R}^d$, $d \in \mathbb{N}$, we denote $\bar{A}$ as the closure of A and ∂A as the boundary of A. If A is Lebesgue-measurable, $|A|$ denotes the Lebesgue measure of A. We indicate a compact subset B of A as $B \subset\subset A$. For any vector $v \in \mathbb{R}^d$, we denote by $\|v\|$ its Euclidean norm.

In many estimations throughout this work, constants appear and we use "C" to denote a generic constant, i. e., C may or may not be the same at different occurrences. In any case, C is always constant with respect to the terms of interest. Moreover, as some lengthy estimates spread over several lines, we often provide explanatory text between different lines of one estimate. In this case, a binary relation like e. g., "$\leq$" at the very beginning of a line indicates that the preceding equality or inequality is being continued.

1.2.1 Spaces on bounded domains

Throughout this work, Ω denotes a bounded polyhedral domain in $\mathbb{R}^d$, $d \in \mathbb{N}$. In the following, we define some well-known spaces to fix notation and for completeness, we review some basic results. Details can be found in e. g., [1, 19, 20, 21, 24, 68].

Definition 1.2.1 (Spaces of differentiable functions)
Let $\alpha \in \mathbb{N}_0^d$ be a multi-index and let $|\alpha| := \sum_{i=1}^d \alpha_i$. For an $|\alpha|$-times continuously differentiable function $f : \Omega \to \mathbb{R}$ we define

$$D^\alpha f := \frac{\partial^{|\alpha|} f}{\partial x_1^{\alpha_1} \dots \partial x_d^{\alpha_d}}, \qquad D^{(0,\dots,0)} f := f.$$

For any $k \in \mathbb{N}_0$ we define the following spaces of continuous and differentiable functions

$$C^k(\Omega) := \{f : \Omega \to \mathbb{R} \mid f \text{ is continuous and}$$
$$k\text{-times cont. differentiable}\},$$
$$C^\infty(\Omega) := \bigcap_{k \in \mathbb{N}_0} C^k(\Omega)$$

as well as

$$C_0^k(\Omega) := \{f \in C^k(\Omega) \mid \operatorname{supp} f \subset\subset \Omega\} \qquad \text{for } k \in \mathbb{N}_0 \cup \{\infty\}.$$

Definition 1.2.2 (Lebesgue spaces)
We define $L_{\mathrm{loc}}^1(\Omega)$ to be the set of locally integrable functions:

$$L_{\mathrm{loc}}^1(\Omega) := \left\{ f : \Omega \to \mathbb{R} \mid f \text{ is measurable and } \int_A f \, dx < \infty \right.$$
$$\left. \text{for all } A \subset\subset \Omega \right\}$$

For $p \in [1, \infty]$ we define

$$L^p(\Omega) := \{f : \Omega \to \mathbb{R} \mid f \text{ is measurable and } \|f\|_{L^p(\Omega)} < \infty\},$$

where the corresponding norms are defined as

$$\|f\|_{L^p(\Omega)} := \begin{cases} \int_\Omega |f(x)|^p \, dx^{1/p} & \text{for } p < \infty, \\ \operatorname{ess\,sup}_{x \in \Omega} |f(x)| & \text{for } p = \infty. \end{cases}$$

As usual, the Lebesgue spaces are actually defined as classes of functions whose values coincide everywhere except for a set of measure zero. With this identification, the Lebesgue spaces $L^p(\Omega)$ equipped with the corresponding norm $\|\cdot\|_{L^p(\Omega)}$ are Banach spaces, which are particularly reflexive if and only if $1 < p < \infty$. In the case of $p = 2$, we may also define a scalar product via

$$\langle f, g \rangle_{L^2(\Omega)} := \int_\Omega f(x)g(x) \, dx \qquad \text{for all } f, g \in L^2(\Omega),$$

giving $L^2(\Omega)$ the additional structure of a Hilbert space. Moreover, throughout this work, we mostly omit the subscript when referring to the $L^2(\Omega)$ norm or scalar product, i.e., we write $\|\cdot\| := \|\cdot\|_{L^2(\Omega)}$ respectively $\langle \cdot\,,\cdot\rangle := \langle \cdot\,,\cdot\rangle_{L^2(\Omega)}$. We only employ the subscript to obviate confusion or to particularly emphasize that we refer to the $L^2(\Omega)$ norm or scalar product.

Definition 1.2.3 (Weak derivatives)
Suppose $f, g \in L^1_{\mathrm{loc}}(\Omega)$ and let $\alpha \in \mathbb{N}_0^d$ be a multi-index. We say that g is the α-th *weak derivative* of f, provided

$$\int_\Omega f D^\alpha \varphi\, dx = (-1)^{|\alpha|} \int_\Omega g\varphi\, dx \qquad \text{for all } \varphi \in C_0^\infty(\Omega).$$

In analogy to the notation for classic derivatives, we write $D^\alpha f := g$ for the α-th weak derivative of f.

Remark 1.2.4 (Weak derivatives)
The concept of weak derivatives truly generalizes the notion of classic derivatives. A weak derivative $D^\alpha f$, if it exists, is uniquely defined up to a set of measure zero. In particular, if $f \in C^1(\Omega)$, i.e., f is differentiable in the classic sense, the classic and weak derivatives coincide. Moreover, well-known concepts involving classical derivatives like e.g., the rules of differentiation or integration by parts, transfer to weak derivatives. A nice summary on weak derivatives can be found in [19, Chapter 5.2] or [20, Chapter 7], for a more thorough discourse we refer to [1].

Definition 1.2.5 (Sobolev spaces)
Let $\alpha \in \mathbb{N}_0^d$ be a multi-index, $k \in \mathbb{N}$ and $p \in [1, \infty]$.

a) We define the Sobolev space

$$W^{k,p}(\Omega) := \{f \in L^p(\Omega) \mid \text{there exist weak derivatives}$$
$$D^\alpha f \in L^p(\Omega) \text{ for all } |\alpha| \le k\}$$

with the norm

$$\|f\|_{W^{k,p}(\Omega)} := \begin{cases} \left(\sum_{|\alpha|\le k} \|D^\alpha f\|_{L^p(\Omega)}^p\right)^{1/p} & \text{for } p < \infty, \\ \max_{|\alpha|\le k} \|D^\alpha f\|_{L^\infty(\Omega)} & \text{for } p = \infty \end{cases}$$

as well as the seminorm

$$|f|_{W^{k,p}(\Omega)} := \begin{cases} \left(\sum_{|\alpha|= k} \|D^\alpha f\|_{L^p(\Omega)}^p\right)^{1/p} & \text{for } p < \infty, \\ \max_{|\alpha|= k} \|D^\alpha f\|_{L^\infty(\Omega)} & \text{for } p = \infty. \end{cases}$$

b) We define the Sobolev space with zero boundary values $W_0^{k,p}(\Omega)$ as the closure of $C_0^\infty(\Omega)$ in $W^{k,p}(\Omega)$.

c) We define $W^{-k,q}(\Omega)$ as the dual space of $W^{k,p}(\Omega)$ where p and q are dual exponents, i. e., $\frac{1}{p} + \frac{1}{q} = 1$. For $f \in W^{-k,q}(\Omega)$ and $g \in W^{k,p}(\Omega)$ we denote by $\langle f, g\rangle_{W^{-k,q}(\Omega) \times W^{k,p}(\Omega)}$ the dual pairing. As a norm on $W^{-k,q}(\Omega)$ we define the standard operator norm

$$\|f\|_{W^{-k,q}(\Omega)} := \sup_{g \in W^{k,p}(\Omega)} \frac{\langle f, g\rangle_{W^{-k,q}(\Omega) \times W^{k,p}(\Omega)}}{\|g\|_{W^{k,p}(\Omega)}}.$$

The Sobolev spaces $W^{k,p}(\Omega)$ equipped with the corresponding norm $\|\cdot\|_{W^{k,p}(\Omega)}$ are Banach spaces, which are particularly reflexive if and only if $1 < p < \infty$. Inheriting the additional structure of $L^2(\Omega)$, the Sobolev spaces $W^{k,2}(\Omega)$ and $W_0^{k,2}(\Omega)$ are also Hilbert spaces and we write

$$H^k(\Omega) := W^{k,2}(\Omega), \qquad H_0^k(\Omega) := W_0^{k,2}(\Omega).$$

In the following chapters, we particularly use those Hilbert spaces for $k = 1$ and we define $H^{-1}(\Omega)$ as the dual space of $H_0^1(\Omega)$. The Poincaré-Friedrichs inequality stated below is particularly important as it allows for using the seminorm $|\cdot|_{H^k}$ as an equivalent norm on $H_0^k(\Omega)$.

Theorem 1.2.6 (Poincaré-Friedrichs inequality)
There is a constant C exclusively depending on the dimension d and the diameter of Ω such that

$$\|f\|_{H^k(\Omega)} \leq C|f|_{H^k(\Omega)} \qquad \text{for all } f \in H_0^k(\Omega).$$

In case of $k = 1$, the seminorm $|\cdot|_{H^1(\Omega)} = (\sum_{|\alpha|=1} \|D^\alpha \cdot\|_{L^2(\Omega)}^2)^{1/2}$ is usually denoted as $\|\nabla \cdot\|_{L^2(\Omega)}$. We assume this notation and hence, in light of Poincaré-Friedrichs inequality, we may define $\|\cdot\|_{H_0^1(\Omega)} := \|\nabla \cdot\|_{L^2(\Omega)}$ as a norm on $H_0^1(\Omega)$. The corresponding scalar product on $H_0^1(\Omega)$, which induces this norm is given by

$$\langle f, g\rangle_{H_0^1(\Omega)} := \langle \nabla f, \nabla g\rangle_{L^2(\Omega)} \qquad \text{for all } f, g \in H_0^1(\Omega).$$

Theorem 1.2.7 (Trace theorem)
There exists a unique linear operator $T : H^1(\Omega) \to L^2(\partial\Omega)$, *the so-called* trace operator, *such that*

$$\|Tf\|_{L^2(\partial\Omega)} \le C \, \|f\|_{H^1(\Omega)} \qquad \text{for all } f \in H^1(\Omega),$$
$$Tf = f \qquad \text{for all } f \in H^1(\Omega) \cap C(\bar{\Omega}),$$

with a constant C exclusively depending on Ω.

Motivated by the trace operator acting as the identity on continuous functions, we conveniently write f instead of Tf. With the notion of traces, the space $H_0^1(\Omega)$ can equivalently be rewritten as

$$H_0^1(\Omega) = \{f \in H^1(\Omega) \mid f = 0 \text{ on } \partial\Omega\}.$$

As a last standard result, we cite Green's formula, which is a consequence of the Gauß Theorem and allows for integration by parts also in $H^1(\Omega)$.

Theorem 1.2.8 (Green's Formula)
Let $f, g \in H^1(\Omega)$ and denote by $\eta(x) = (\eta_1(x), \ldots, \eta_d(x))^\mathsf{T}$ the outer unit normal of $\partial\Omega$ in x. Then for any $i = 1, \ldots, d$, Green's formula

$$\int_\Omega \partial_i f \, g \, dx = -\int_\Omega f \, \partial_i g \, dx + \int_{\partial\Omega} fg \, \eta_i \, ds$$

holds. Equivalently, if $f \in H^1(\Omega)$ and $g \in H^1(\Omega; \mathbb{R}^d)$, there holds

$$\int_\Omega f \operatorname{div} g \, dx = -\int_\Omega g \cdot \nabla f \, dx + \int_{\partial\Omega} fg \cdot \eta \, ds.$$

1.2.2 Spaces involving time

For the time dependent problems in this work, we consider a bounded time interval $I \subset \mathbb{R}$ next to the spacial domain Ω. It is convenient to consider the involved space and time dependent functions not as functions

$$f : \Omega \times I \to \mathbb{R}$$

mapping a point $x \in \Omega$ and a time $t \in I$ to a value $f(x, t) \in \mathbb{R}$, but rather as a function

$$f : I \to H^1(\Omega)$$

mapping a time $t \in I$ to a *function* in space. As a foundation for the upcoming definition of suitable time dependent spaces, we state a brief introduction to integration of Banach space valued functions, for details we refer to [18, Chapter 18], [69, Chapter V.5], or [68, §24].

Definition 1.2.9 (Bochner integral)

Let $-\infty < a < b < \infty$ and let $(\mathbb{X}, \|\cdot\|_{\mathbb{X}})$ be a real Banach space.

a) A function $s : [a, b] \to \mathbb{X}$ is called *simple* if there are $\varphi_1, \ldots, \varphi_N \in \mathbb{X}$ and Lebesgue-measurable disjoint sets $\mathcal{M}_1, \ldots, \mathcal{M}_N \subset [a, b]$ with $|\mathcal{M}_i| < \infty$, $i = 1, \ldots, N$, such that

$$s(t) = \sum_{i=1}^{N} \chi_{\mathcal{M}_i}(t)\, \varphi_i \qquad t \in [a, b],$$

where $\chi_{\mathcal{M}_i}$ denotes the indicator function of the set $\mathcal{M}_i \subset [a, b]$.

b) For a simple function $s(t) = \sum_{i=1}^{N} \chi_{\mathcal{M}_i}(t)\, \varphi_i$ we define

$$\int_a^b s(t)\, dt := \sum_{i=1}^{N} |\mathcal{M}_i|\, \varphi_i \in \mathbb{X}.$$

c) A function $f : [a, b] \to \mathbb{X}$ is *Bochner-measurable* if there exists a sequence of simple functions $\{s_k\}_{k \in \mathbb{N}}$ such that

$$\lim_{k \to \infty} s_k(t) = f(t) \qquad \text{for almost all } t \in [a, b].$$

d) Beyond that, a Bochner-measurable function $f : [a, b] \to \mathbb{X}$ is *Bochner-summable* if

$$\lim_{k \to \infty} \int_a^b \|s_k(t) - f(t)\|_{\mathbb{X}}\, dt = 0.$$

e) For a Bochner-summable function $f : [a, b] \to \mathbb{X}$ we define

$$\int_a^b f(t)\, dt := \lim_{k \to \infty} \int_a^b s_k(t)\, dt.$$

A straightforward calculation shows that the sequence given by $\{\int_a^b s_k(t)\, dt\}_{k\in\mathbb{N}}$ is a Cauchy sequence in the Banach space $\mathbb{X}$ and hence, the limit $\lim_{k\to\infty}\int_a^b s_k(t)\, dt$ in above definition exists. Moreover, it is independent of the actual choice of the sequence $\{s_k\}_{k\in\mathbb{N}}$ and thus, the Bochner integral $\int_a^b f(t)\, dt$ with $f : [a, b] \to \mathbb{X}$ is well-defined.

Theorem 1.2.10
A Bochner-measurable function $f : [a, b] \to \mathbb{X}$ is Bochner-summable if and only if $t \mapsto \|f(t)\|_{\mathbb{X}}$ is Lebesgue-summable. In this case

$$\left\| \int_a^b f(t)\, dt \right\|_{\mathbb{X}} \leq \int_a^b \|f(t)\|_{\mathbb{X}}\, dt.$$

Definition 1.2.11 (Spaces of Banach space valued functions)
For $-\infty < a < b < \infty$, $p \in [1, \infty]$ and a Banach space $(\mathbb{X}, \|\cdot\|_{\mathbb{X}})$ we define:

a) The Lebesgue space

$$L^p(a, b; \mathbb{X}) := \{f : [a, b] \to \mathbb{X} \,|\, f \text{ is Bochner-measurable and}$$
$$\|f\|_{L^p(a,b;\mathbb{X})} < \infty\},$$

where the corresponding norms are defined as

$$\|f\|_{L^p(a,b;\mathbb{X})} := \begin{cases} \left(\int_a^b \|f(t)\|_{\mathbb{X}}^p \, dt\right)^{1/p} & \text{for } p < \infty, \\ \operatorname{ess\,sup}_{t\in[a,b]} \|f(t)\|_{\mathbb{X}} & \text{for } p = \infty. \end{cases}$$

b) The space of continuous functions

$$C(a, b; \mathbb{X}) := \{f : [a, b] \to \mathbb{X} \,|\, f \text{ is continuous and}$$
$$\|f\|_{C(a,b;\mathbb{X})} < \infty\},$$

where the norm is defined as

$$\|f\|_{C(a,b;\mathbb{X})} := \max_{t\in[a,b]} \|f(t)\|_{\mathbb{X}}.$$

The spaces $L^p(a, b; \mathbb{X})$ and $C(a, b; \mathbb{X})$ equipped with their respective norm $\|\cdot\|_{L^p(a,b;\mathbb{X})}$ or $\|\cdot\|_{C(a,b;\mathbb{X})}$ are Banach spaces and Hölder's inequality holds in the Lebesgue space $L^p(a, b; \mathbb{X})$.

Theorem 1.2.12 (Hölder's inequality)
Let $\mathbb{X}$ be a real Banach space and denote its dual by $\mathbb{X}'$. Moreover, for $1 \leq p \leq \infty$ let q be the dual exponent to p, i. e., $\frac{1}{p} + \frac{1}{q} = 1$. Then, for $f \in L^p(a, b; \mathbb{X})$ and $g \in L^q(a, b; \mathbb{X}')$, we have $\langle g(\cdot), f(\cdot) \rangle_{\mathbb{X}' \times \mathbb{X}} \in L^1(a, b)$ and Hölder's inequality holds, i. e.,

$$\int_a^b \langle g(t), f(t) \rangle_{\mathbb{X}' \times \mathbb{X}} \, dt \leq \|g\|_{L^q(a,b;\mathbb{X}')} \|f\|_{L^p(a,b;\mathbb{X})}.$$

In the special case $p = 2$, Hölder's inequality becomes the Cauchy-Schwarz inequality. Moreover, the notion of weak derivatives generalizes straightforwardly to the case of Banach space valued functions:

Definition 1.2.13 (Weak derivative)
Let $(\mathbb{X}, \|\cdot\|_{\mathbb{X}})$ be a real Banach space and suppose $f, g \in L^1(a, b; \mathbb{X})$. We say that g is the weak derivative of f, provided

$$\int_a^b f(t)\, \varphi'(t) \, dt = - \int_a^b g(t)\, \varphi(t) \, dt \qquad \text{for all } \varphi \in C_0^\infty(a, b).$$

For the weak derivative we write $f' := g$.

With this notion of weak derivatives, also its properties and the definition of Sobolev spaces transfers straightforwardly to the situation of Banach space valued functions. In particular, we define

$$H^1(0, T; L^2(\Omega)) := \{ f \in L^2(0, T; L^2(\Omega)) \mid \text{the weak derivative}$$
$$f' \in L^2(0, T; L^2(\Omega)) \text{ exists}\},$$

which is used for prescribing the regularity of the right hand side of the PDE introduced in the next chapter.

Chapter 2

Analysis of parabolic PDEs

The class of linear parabolic partial differential equations, which is subject of this work is given by

$$\partial_t u(x,t) + \mathcal{L}u(x,t) = f(x,t) \quad \text{in } \Omega \times (0,T),$$
$$u(x,t) = 0 \quad \text{on } \partial\Omega \times (0,T), \qquad (2.1)$$
$$u(\cdot,0) = u_0(x) \quad \text{in } \Omega.$$

Here, ∂_t denotes the partial derivative with respect to time and u is the unknown depending on space and time. By $\mathcal{L}$ we denote a second order differential operator which is introduced shortly. In the simplest case of $\mathcal{L} = -\Delta$, problem (2.1) becomes the heat equation.

Since it will greatly clarify the subsequent analysis, we use a special interpretation of the time and space dependent functions u and f. Particularly, we regard u not as a function

$$u : \Omega \times (0,T) \to \mathbb{R}$$

mapping from a space-time-domain to the reals but rather as a function

$$u : (0,T) \to H_0^1(\Omega)$$

mapping a time t to a (space dependent) *function* $u(t) \in H_0^1(\Omega)$, cf. Section 1.2.2. In this sense, we identify

$$(u(t))(x) := u(x,t) \qquad \text{for all } x \in \Omega, \, t \in [0,T].$$

Note that with this identification, we no longer consider the *partial* derivative $\partial_t u(x,t)$ but rather the derivative $\frac{d}{dt}u(t)$ of a function depending on only *one* variable. We denote this temporal derivative as $u'(t) := \frac{d}{dt}u(t)$.

We understand the data f accordingly as a function mapping a time t to a function $f(t) \in L^2(\Omega)$. Recalling the time dependent spaces from Definition 1.2.11, we particularly demand

$$f \in H^1(0,T;L^2(\Omega)).$$

Moreover, we demand

$$u_0 \in L^2(\Omega)$$

and hence, the initial condition in equation (2.1) is to be understood in an L^2-sense. In this setting, we introduce the linear operator $\mathcal{L}$ as second order elliptic operator with respect to space, which is represented for $v \in H^1(\Omega)$ via

$$\mathcal{L}\, v := -\operatorname{div}(\mathbf{A}\nabla v) + \mathbf{b}\cdot\nabla v + c\, v. \tag{2.2}$$

Here, $\mathbf{A} \in L^\infty(\Omega; \mathbb{R}^{d\times d})$ is a matrix valued function with $\mathbf{A}(x)$ being symmetric positive definite with eigenvalues in $0 < a_- \leq a_+ < \infty$, i.e.,

$$a_- \|\xi\|^2 \leq \mathbf{A}(x)\xi \cdot \xi \leq a_+ \|\xi\|^2 \quad \text{for all } \xi \in \mathbb{R}^d, \, x \in \Omega, \tag{2.3a}$$

and piecewise Lipschitz over some initial triangulation $\mathcal{G}_{\text{init}}$, see Section 3.1. Moreover, $\mathbf{b} \in L^\infty(\Omega; \mathbb{R}^d)$ is a divergence free vector field, i.e.,

$$\operatorname{div} \mathbf{b}(x) = 0 \qquad \text{for all } x \in \Omega, \tag{2.3b}$$

and $c \in L^\infty(\Omega)$ is non-negative, i.e.,

$$c(x) \geq 0 \qquad \text{for all } x \in \Omega. \tag{2.3c}$$

For the weak formulation, the conditions on $\mathbf{A}$ and f can be re-laxed: The piecewise Lipschitz continuity of $\mathbf{A}$ is only needed for the residual error estimator in Chapter 3. Regarding f, the convergence proof in Chapter 4 uses the additional regularity with respect to time, otherwise it is sufficient to demand $f \in L^2(0, T; L^2(\Omega))$.

In the following, we derive the weak formulation of problem (2.1) in Section 2.1 before establishing existence of a unique solution in Section 2.2. In Section 2.3, we finally present a stability estimate, which is closely related to a uniform energy estimate which plays an essential role in Chapter 4.

2.1 Weak formulation

As a foundation for the upcoming analysis as well as the discretization and the computational approach presented in Chapters 3 and 4, we will introduce the weak formulation of problem (2.1) in this section.

To get a notion of a suitable solution space, we consider a fixed time $t \in [0, T]$ and assume $u(t) \in H_0^1(\Omega)$. Following the standard procedure for deriving a weak formulation, we multiply (2.1) with an arbitrary test function $v \in H_0^1(\Omega)$ and integrate over the domain Ω. Applying Green's formula (see Theorem 1.2.8) to the second order term (implicitly assuming $\mathbf{A}\nabla u(t) \in H^1(\Omega; \mathbb{R}^d)$), we have

$$\int_\Omega u'(t)\,v + \mathbf{A}\nabla u(t) \cdot \nabla v + \mathbf{b} \cdot \nabla u(t)\,v + c\,u\,v\,dx = \int_\Omega f(t)\,v\,dx, \quad (2.4)$$

for all $v \in H_0^1(\Omega)$, $t \in (0, T)$, since (the trace of) v vanishes on $\partial\Omega$. We observe that equation (2.4) itself does *not* require above H^1-regularity of $\mathbf{A}\nabla u(t)$ in order to be well-posed. Equation (2.4) motivates the introduction of the bilinear form $\mathcal{B}$.

Definition 2.1.1 (Bilinear form)
We define the bilinear form

$$\mathcal{B} : H_0^1(\Omega) \times H_0^1(\Omega) \to \mathbb{R}$$

$$\mathcal{B}(v\,,\,w) := \int_\Omega \mathbf{A}\nabla v \cdot \nabla w + \mathbf{b} \cdot \nabla v\,w + c\,v\,w\,dx.$$

In the following, we briefly focus on the first order term. To that end, we conveniently abbreviate for all $v, w \in H_0^1(\Omega)$

$$a_{\mathbf{b}}(v, w) := \int_\Omega \mathbf{b} \cdot \nabla v \, w \, dx \tag{2.5}$$

introducing the auxiliary bilinear form $a_{\mathbf{b}} : H_0^1(\Omega) \times H_0^1(\Omega) \to \mathbb{R}$. We show that $a_{\mathbf{b}}(\cdot, \cdot)$ is anti-symmetric by first noting that for $v, w \in H_0^1(\Omega)$ it holds

$$\operatorname{div}(\mathbf{b}\, v\, w) = \mathbf{b} \cdot \nabla v \, w + v \operatorname{div}(\mathbf{b}\, w) = \mathbf{b} \cdot \nabla v \, w + v \operatorname{div} \mathbf{b} \, w + v \, \mathbf{b} \cdot \nabla w.$$

Exploiting the assumption that $\mathbf{b}$ is divergence free by (2.3b), this yields

$$a_{\mathbf{b}}(v, w) = \int_\Omega \operatorname{div}(\mathbf{b}\, v\, w) - v \, \mathbf{b} \cdot \nabla w \, dx$$

$$= -\int_\Omega v \, \mathbf{b} \cdot \nabla w \, dx = -a_{\mathbf{b}}(w, v), \tag{2.6}$$

where we also used Green's formula and the fact that $v = 0$ on $\partial\Omega$. Hence, we see that $a_{\mathbf{b}}(\cdot, \cdot)$ is anti-symmetric which particularly implies

$$a_{\mathbf{b}}(v, v) = 0 \qquad \text{for all } v \in H_0^1(\Omega). \tag{2.7}$$

Considering this, we realize that $\mathcal{B}(v, v) = \int_\Omega \mathbf{A}\nabla v \cdot \nabla v + c\, v^2 \, dx$ does not involve the first order term. This allows for the following definition of the energy norm, where we particularly recall $H^{-1}(\Omega)$ to be the dual space of $H_0^1(\Omega)$ from Section 1.2.

Definition 2.1.2 (Energy norm)
We define

a) for $v \in H_0^1(\Omega)$ the *energy*-norm

$$\|v\|_\Omega := \mathcal{B}(v, v)^{1/2} = \left(\int_\Omega \mathbf{A}\nabla v \cdot \nabla v + c\, v^2 \, dx \right)^{1/2};$$

b) for $v \in H^{-1}(\Omega)$ the norm $\|v\|_*$ as the operator norm

$$\|v\|_* := \sup_{w \in H_0^1(\Omega)} \frac{\langle v, w \rangle_{H^{-1}(\Omega) \times H_0^1(\Omega)}}{\|w\|_\Omega}.$$

The following lemma justifies the above definition of the energy norm and relates it to the standard norm $\|\cdot\|_{H_0^1(\Omega)}$ on $H_0^1(\Omega)$.

Lemma 2.1.3 (Properties of $\mathcal{B}(\cdot\,,\,\cdot)$ and the energy norm)
Let $\mathcal{B}(\cdot\,,\,\cdot)$ and $\|\cdot\|_\Omega$ denote the bilinear form respectively the energy norm as introduced in Definitions 2.1.1 and 2.1.2.

a) The energy norm $\|\cdot\|_\Omega$ is a norm on $H_0^1(\Omega)$ which is equivalent to the norm $\|\cdot\|_{H_0^1(\Omega)} = \|\nabla\cdot\|_{L^2(\Omega)}$ on $H_0^1(\Omega)$, i. e.,

$$\mathcal{C}_- \|v\|_{H_0^1(\Omega)} \leq \|v\|_\Omega \leq \mathcal{C}_+ \|v\|_{H_0^1(\Omega)} \qquad \text{for all } v \in H_0^1(\Omega).$$

The constant $\mathcal{C}_-$ only depends on the smallest eigenvalue a_- of $\mathbf{A}$, whereas $\mathcal{C}_+$ only depends on the biggest eigenvalue a_+ of $\mathbf{A}$, the zero order coefficient c as well as the domain Ω and the dimension d.

b) The Poincaré-Friedrichs inequality

$$\|v\|_{L^2(\Omega)} \leq \mathcal{C}_\mathcal{P} \|v\|_\Omega \qquad \text{for all } v \in H_0^1(\Omega)$$

holds with a constant $\mathcal{C}_\mathcal{P}$ only depending on the smallest eigenvalue a_- of $\mathbf{A}$, the domain Ω and the dimension d.

c) The bilinear form $\mathcal{B}(\cdot\,,\,\cdot)$ is continuous with respect to the energy norm. More precisely, there is a constant $\mathcal{C}_\mathcal{B}$ such that

$$|\mathcal{B}(v\,,\,w)| \leq \mathcal{C}_\mathcal{B} \|v\|_\Omega \|w\|_\Omega.$$

The constant $\mathcal{C}_\mathcal{B}$ is given by $\mathcal{C}_\mathcal{B} = 1 + \|\mathbf{b}\|_{L^\infty(\Omega)} \frac{\mathcal{C}_\mathcal{P}}{\mathcal{C}_-}$ and hence depends on a_-, the first order coefficient $\mathbf{b}$ as well as the domain Ω and the dimension d.

Proof. Let $v, w \in H_0^1(\Omega)$ be arbitrary throughout this proof. We start proving
a) From the definition of $\|\cdot\|_\Omega$, we directly see that

$$\|\alpha\, v\|_\Omega = |\alpha| \, \|v\|_\Omega \qquad \text{for all } \alpha \in \mathbb{R}$$

and that the triangle inequality holds,

$$\|v + w\|_\Omega \leq \|v\|_\Omega + \|w\|_\Omega.$$

With the properties of $\mathbf{A}$ and c from (2.3), we estimate from the definition of the energy norm

$$\|v\|_{\Omega}^2 \leq a_+ \|\nabla v\|_{L^2(\Omega)}^2 + \|c\|_{L^\infty(\Omega)} \|v\|_{L^2(\Omega)}^2$$

and by Poincaré-Friedrichs inequality (see Theorem 1.2.6), we deduce with $\tilde{\mathcal{C}}_{\mathcal{P}}$ denoting the Poincaré constant of the estimate $\|v\| \leq \tilde{\mathcal{C}}_{\mathcal{P}} \|\nabla v\|$

$$\leq \max\{a_+, \tilde{\mathcal{C}}_{\mathcal{P}}^2 \|c\|_{L^\infty(\Omega)}\} \|\nabla v\|_{L^2(\Omega)}^2 = \mathcal{C}_+^2 \|v\|_{H_0^1(\Omega)}^2 , \quad (2.8)$$

where we defined $\mathcal{C}_+^2 := \max\{a_+, \tilde{\mathcal{C}}_{\mathcal{P}}^2 \|c\|_{L^\infty(\Omega)}\}$. Note that the Poincaré constant $\tilde{\mathcal{C}}_{\mathcal{P}}$ depends on the domain Ω and the dimension d. On the other hand, the definition of the energy norm in combination with the properties of $\mathbf{A}$ and c from (2.3) yields

$$\|v\|_{\Omega}^2 \geq a_- \|\nabla v\|_{L^2(\Omega)}^2 = \mathcal{C}_-^2 \|v\|_{H_0^1(\Omega)}^2 , \quad (2.9)$$

where we defined $\mathcal{C}_-^2 := a_-$. From (2.8) and (2.9) we additionally conclude

$$\|v\|_{\Omega} = 0 \quad \Leftrightarrow \quad v = 0 \quad \text{in } H_0^1(\Omega)$$

and hence, the energy norm $\|\cdot\|_{\Omega}$ is indeed a norm on $H_0^1(\Omega)$.

b) Using the Poincaré-Friedrichs inequality on $H_0^1(\Omega)$, we estimate

$$\|v\|_{L^2(\Omega)} \leq \tilde{\mathcal{C}}_{\mathcal{P}} \|v\|_{H_0^1(\Omega)} \leq \frac{\tilde{\mathcal{C}}_{\mathcal{P}}}{\mathcal{C}_-} \|v\|_{\Omega} \quad (2.10)$$

and we define $\mathcal{C}_{\mathcal{P}} := \tilde{\mathcal{C}}_{\mathcal{P}}/\mathcal{C}_-$.

c) To establish continuity, we conveniently split $\mathcal{B}(\cdot, \cdot)$ into

$$\mathcal{B}(v, w) = a_{\mathbf{A}c}(v, w) + a_{\mathbf{b}}(v, w) \quad (2.11)$$

with auxiliary bilinear forms

$$a_{\mathbf{A}c}(v, w) := \int_{\Omega} \mathbf{A}\nabla v \cdot \nabla w + c\, v\, w\, dx,$$

$$\text{and} \quad a_{\mathbf{b}}(v, w) := \int_{\Omega} \mathbf{b} \cdot \nabla v\, w\, dx \quad \text{as in (2.5).}$$

We estimate $a_{\mathbf{A}c}$ and $a_{\mathbf{b}}$ individually. Since $\mathbf{A}$ is symmetric positive definite and c is non-negative, we directly see that $a_{\mathbf{A}c}(\cdot,\cdot)$ is a scalar product. Hence, we may employ Cauchy-Schwarz inequality to estimate

$$|a_{\mathbf{A}c}(v,w)| \leq a_{\mathbf{A}c}(v,v)^{1/2}\, a_{\mathbf{A}c}(w,w)^{1/2} = \|v\|_{\Omega}\, \|w\|_{\Omega}\,, \qquad (2.12)$$

where we also used the definitions of $a_{\mathbf{A}c}$ and the energy norm. We estimate the remaining term $a_{\mathbf{b}}$ also using Cauchy-Schwarz inequality and obtain

$$|a_{\mathbf{b}}(v,w)| = \left|\int_{\Omega} \mathbf{b}\cdot\nabla v\, w\, dx\right| \leq \|\mathbf{b}\cdot\nabla v\|_{L^2(\Omega)}\, \|w\|_{L^2(\Omega)}$$
$$\leq \|\mathbf{b}\|_{L^\infty(\Omega)}\, \|\nabla v\|_{L^2(\Omega)}\, \|w\|_{L^2(\Omega)}\,.$$

Employing (2.9) to estimate $\|\nabla v\|_{L^2(\Omega)} \leq \mathcal{C}_{-}^{-1}\, \|v\|_{\Omega}$ and using Poincaré-Friedrichs inequality (2.10) to estimate $\|w\|_{L^2(\Omega)}$ we derive

$$\leq \|\mathbf{b}\|_{L^\infty(\Omega)}\, \frac{\mathcal{C}_{\mathcal{P}}}{\mathcal{C}_{-}}\, \|v\|_{\Omega}\, \|w\|_{\Omega}\,. \qquad (2.13)$$

Substituting estimates (2.12) and (2.13) into the splitting (2.11), we derive

$$|\mathcal{B}(v,w)| \leq \left(1 + \|\mathbf{b}\|_{L^\infty(\Omega)}\, \frac{\mathcal{C}_{\mathcal{P}}}{\mathcal{C}_{-}}\right)\, \|v\|_{\Omega}\, \|w\|_{\Omega} =: \mathcal{C}_{\mathcal{B}}\, \|v\|_{\Omega}\, \|w\|_{\Omega}\,.$$

$\square$

Remark 2.1.4 (Cauchy-Schwarz inequality)
In the foregoing lemma, we showed that the bilinear form $\mathcal{B}(\cdot,\cdot)$ is continuous with a constant $\mathcal{C}_{\mathcal{B}}$, i.e.,

$$|\mathcal{B}(v,w)| \leq \mathcal{C}_{\mathcal{B}}\, \|v\|_{\Omega}\, \|w\|_{\Omega}\,. \qquad (2.14)$$

In the case of $\mathcal{B}(\cdot,\cdot)$ being symmetric, which holds if and only if $\mathbf{b} = 0$, this bilinear form defines a scalar product with induced norm $\|\cdot\|_{\Omega}$. In this case, estimate (2.14) is the Cauchy-Schwarz inequality and we have $\mathcal{C}_{\mathcal{B}} = 1$, which also follows from the definition of $\mathcal{C}_{\mathcal{B}}$ in above proof.

Inspired by Lemma 2.1.3, we mostly use the energy norm $\|\cdot\|_\Omega$ from Definition 2.1.2 as a norm on $H_0^1(\Omega)$. Whenever referring to the standard norm $\|\cdot\|_{H_0^1(\Omega)} = \|\nabla\cdot\|_{L^2(\Omega)}$ on $H_0^1(\Omega)$, we explicitly write $\|\cdot\|_{H_0^1(\Omega)}$. However, we conveniently define the $L^2 - H_0^1$ norm exclusively with respect to the energy norm:

Definition 2.1.5 (Norm on $L^2(0,T;H_0^1(\Omega))$)
For $-\infty < a < b < \infty$, we define

$$\|v\|_{L^2(a,b;H_0^1(\Omega))} := \left(\int_a^b \|v(t)\|_\Omega^2 \, dt \right)^{1/2}$$

for all $v \in L^2(a,b;H_0^1(\Omega))$.

Remark 2.1.6 (Dual space)
Since we equipped the Hilbert space $H_0^1(\Omega)$ with the alternate norm $\|\cdot\|_\Omega$, it is not clear whether the corresponding dual space coincides with the dual space of the Hilbert space $(H_0^1(\Omega), \|\cdot\|_{H_0^1(\Omega)})$. However, since the two employed norms $\|\cdot\|_\Omega$ and $\|\cdot\|_{H_0^1(\Omega)}$ are equivalent (cf. Lemma 2.1.3), a functional which is bounded with respect to $\|\cdot\|_\Omega$ is also bounded with respect to $\|\cdot\|_{H_0^1(\Omega)}$ and vice versa. Thus, the Hilbert spaces $(H_0^1(\Omega), \|\cdot\|_\Omega)$ and $(H_0^1(\Omega), \|\cdot\|_{H_0^1(\Omega)})$ do possess the same dual space $H^{-1}(\Omega)$.

In order to identify a suitable solution space for (2.1), we employ the simplest possible choice of parameters in (2.4), i.e., $\mathbf{A} \equiv \mathrm{id}$, $\mathbf{b} = 0$, $c = 0$ and $f = 0$, which produces the variational formulation of the heat equation from (2.4):

$$\langle u'(t), v \rangle = -\langle \nabla u(t), \nabla v \rangle \qquad \text{for all } v \in H_0^1(\Omega),\ t \in (0,T)$$

Regarding this equation as a definition $\langle u'(t), v \rangle := -\langle \nabla u(t), \nabla v \rangle$, we observe that for $u(t) \in H_0^1(\Omega)$, in general the right hand side $\langle \nabla u(t), \nabla v \rangle$ only defines a functional in $H^{-1}(\Omega)$. Hence, we can not expect $u'(t) \in H_0^1(\Omega)$ but only $u'(t) \in H^{-1}(\Omega)$.

With this motivation, we turn back to the general situation of equation (2.4). Understanding the term $\int_\Omega u'(t)\,v\,dx$ in (2.4) as a *dual pairing* between $H^{-1}(\Omega)$ and $H_0^1(\Omega)$, we conclude for any $v \in H_0^1(\Omega)$

$$\langle u'(t), v \rangle_{H^{-1}(\Omega) \times H_0^1(\Omega)} = \langle f(t), v \rangle - \mathcal{B}(u(t), v),$$

which can be estimated using Cauchy-Schwarz inequality, the continuity of $\mathcal{B}(\cdot\,,\cdot)$, and Poincaré-Friedrichs inequality (cf. Lemma 2.1.3) by

$$\leq \|f(t)\| \, \|v\| + C_{\mathcal{B}} \, \|u(t)\|_{\Omega} \, \|v\|_{\Omega} \leq C(\|f(t)\| + \|u(t)\|_{\Omega}) \, \|v\|_{\Omega}\,.$$

Hence, we derived the bound

$$\|u'(t)\|_{H^{-1}(\Omega)} \leq C(\|f(t)\| + \|u(t)\|_{\Omega}), \qquad (2.15)$$

which verifies above expectation $u' \in H^{-1}(\Omega)$. Assuming square integrability in time, this motivates the following definition of the solution space for parabolic problems.

Definition 2.1.7 (Solution space)
We define the Banach space

$$\mathbb{W}(0,T) := \{u \in L^2(0,T;H^1_0(\Omega)) \mid u' \in L^2(0,T;H^{-1}(\Omega))\}$$

with norm

$$\|u\|_{\mathbb{W}(0,T)} := \left(\int_0^T \|u'\|_*^2 + \|u\|_{\Omega}^2 \, dt \right)^{1/2}.$$

We understand the spaces $L^2(0,T;H^1_0(\Omega))$ and $L^2(0,T;H^{-1}(\Omega))$ as in Definition 1.2.11.

Further investigation of this space is now in order. Particularly, we note that for a function $u \in \mathbb{W}(0,T)$, the derivative $u' \in H^{-1}(\Omega)$ is contained in a different space than the function u itself. To resolve this, we might employ the Riesz isomorphism $\mathcal{R}$ from the Riesz Representation Theorem to identify the two spaces

$$H^1_0(\Omega) \longleftrightarrow_{\mathcal{R}} H^{-1}(\Omega).$$

However, we will *not* use this identification since it employs the scalar product on $H^1_0(\Omega)$ for representing functionals from $H^{-1}(\Omega)$. Opposed to that, we want to use the $L^2(\Omega)$ scalar product for this cause, compare also (2.4). In order to identify $H^1_0(\Omega)$ and $H^{-1}(\Omega)$, we therefore use the concept of Gelfand triples, for details, we refer to [68]. Since $H^1_0(\Omega)$ is continuously and densely embedded in $L^2(\Omega)$,

$$H^1_0(\Omega) \hookrightarrow_i L^2(\Omega),$$

we also have the continuous and dense dual embedding

$$(L^2(\Omega))' \hookrightarrow_{i'} H^{-1}(\Omega).$$

Using the Riesz Representation Theorem to identify $L^2(\Omega)$ with its dual, we then produce the Gelfand triple

$$H_0^1(\Omega) \hookrightarrow_i L^2(\Omega) \hookrightarrow_{i'} H^{-1}(\Omega).$$

In particular, we embedded $H_0^1(\Omega) \hookrightarrow_{i' \circ i} H^{-1}(\Omega)$ without directly using the Riesz Representation Theorem for those two spaces. However, since we used the Riesz Representation Theorem to identify $L^2(\Omega)$ and its dual, we must obey for $f \in L^2(\Omega) \subset H^{-1}(\Omega)$ and $g \in H_0^1(\Omega) \subset L^2(\Omega)$ the compatibility condition

$$\langle f, g \rangle_{H^{-1}(\Omega) \times H_0^1(\Omega)} = \langle f, g \rangle_{L^2(\Omega)}.$$

From this, we see that the identification of $H^{-1}(\Omega)$ and $H_0^1(\Omega)$ via $i' \circ i$ is indeed different from the direct identification via the Riesz Representation Theorem, which employs the scalar product on $H_0^1(\Omega)$.

Whereas these observations put us in the position to state the weak formulation of the actual equation in (2.1), the following theorem is needed for the initial condition to be well-posed, cf. Remarks 2.1.10 and 2.1.11.

Theorem 2.1.8 (Properties of $\mathbb{W}(0,T)$)
The space $\mathbb{W}(0,T)$ continuously embeds into $C([0,T]; L^2(\Omega))$. In particular, for $u \in \mathbb{W}(0,T)$, it holds

$$u \in C([0,T]; L^2(\Omega))$$

after possibly being redefined on a set of measure zero and it holds the estimate

$$\max_{t \in [0,T]} \|u(t)\| \leq C \|u\|_{\mathbb{W}(0,T)}$$

with a constant only depending on T.

Proof. [19, Chapter 5.9.2, Theorem 3]   □

We are now in the position to state the weak formulation of (2.1).

Definition 2.1.9 (Weak solution)
A function $u \in \mathbb{W}(0,T)$ is called a *weak solution* of problem (2.1)
if

$$\langle u'(t), v \rangle + \mathcal{B}(u(t), v) = \langle f(t), v \rangle \quad \text{for all } v \in H_0^1(\Omega),$$
$$\text{f.a.e. } t \in (0,T) \quad \text{(2.16a)}$$
$$u(0) = u_0 \quad \text{(2.16b)}$$

As we will see in Theorems 2.2.5 and 2.2.6, there exists a unique weak solution of problem (2.1). Except from Section 3.3, u will always denote this weak solution throughout this thesis.

Remark 2.1.10 (Representation of u')
From $u \in \mathbb{W}(0,T)$ we only get $u'(t) \in H^{-1}(\Omega)$ (for a. e. $t \in (0,T)$) and consequently, the dual pairing

$$\langle u'(t), v \rangle_{H^{-1}(\Omega) \times H_0^1(\Omega)}$$

should be used in (2.16). For $u'(t) \in L^2(\Omega)$ (for a. e. $t \in (0,T)$), this dual pairing on the other hand can be represented via the Gelfand triple as L^2-scalar product which we denote by $\langle u'(t), v \rangle$. Most importantly, this coincides with the term containing the temporal derivative in equation (2.4).

Remark 2.1.11 (Initial condition)
Even though $u \in \mathbb{W}(0,T)$ only implies that u has L^2-regularity with respect to time, the initial condition $u(0) = u_0$ is well-posed since we have the additional regularity $u \in C(0,T;L^2(\Omega))$ due to Theorem 2.1.8. Nevertheless, with respect to *space* we have to understand the initial condition in the L^2 almost everywhere sense, i. e., $u(0) = u_0$ a. e. in Ω.

Remark 2.1.12 (Relation of weak and strong solutions)
Suppose that u is a classical solution of (2.1). This particularly implies that u possesses sufficient regularity such that the differential operators ∂_t and $\mathcal{L}$ may be applied and the boundary and initial conditions are well-posed. We thus have $u \in C^1(0,T;C^2(\Omega))$ from the differential operators, and $u \in C([0,T];C(\bar{\Omega}))$ from the boundary and initial conditions. Multiplying the equation

$$\partial_t u(x,t) + \mathcal{L}u(x,t) = f(x,t)$$

with an arbitrary $v \in H_0^1(\Omega)$, integrating over Ω and integrating by parts, we deduce

$$\langle \partial_t u(x,t),\, v \rangle + \mathcal{B}(u(x,t),\, v) = \langle f(t),\, v \rangle$$

for all $t \in (0, T)$. We conclude, that u is also a weak solution. Contrary, assume that u is a weak solution with the additional regularity as stated for the classical solution above. We then have

$$\langle u'(t),\, v \rangle + \mathcal{B}(u(t),\, v) = \langle f(t),\, v \rangle$$

for any $v \in H_0^1(\Omega)$ and $u(0) = u_0$. Exploiting the additional regularity, we may particularly revert the integration by parts to obtain

$$\int_\Omega \partial_t u(x,t)\, v + \mathcal{L}u(x,t)\, v \, dx = \int_\Omega f(x,t)\, v\, dx$$

for all $v \in H_0^1(\Omega)$. By the Fundamental Lemma of calculus of variations, we deduce

$$\partial_t u(x,t) + \mathcal{L}u(x,t) = f(x,t).$$

The fact that u is in $H_0^1(\Omega)$ in combination with the additional regularity implies that also the boundary condition of (2.1) is satisfied. Hence, provided sufficient regularity, a weak solution is indeed a strong solution.

2.2   Existence and uniqueness of a solution

In this section, we intend to construct a weak solution u of (2.1), i.e., a function $u \in \mathbb{W}(0, T)$ satisfying (2.16). To that end, we employ the so-called Galerkin method, which first generates approximations to u in certain finite-dimensional spaces. Choosing those spaces appropriately and passing to limits eventually yields the exact solution u.

In order to generate such finite dimensional spaces, we assume that we have at hand a sequence of functions $\{w_k\}_{k\in\mathbb{N}} \subset H_0^1(\Omega)$ such that the set $\{w_k\}_{k\in\mathbb{N}} \subset H_0^1(\Omega)$ is

a) an orthogonal basis of $H_0^1(\Omega)$, i.e.,

$$\langle w_n, w_m \rangle_{H_0^1(\Omega)} = 0 \qquad \text{for } n \neq m;$$

b) an orthonormal basis of $L^2(\Omega)$, i.e.,

$$\langle w_n, w_m \rangle_{L^2(\Omega)} = \delta_{nm},$$

where δ_{nm} denotes the Kronecker symbol.

Such a sequence of functions exists, a possible choice is the (normalized) eigenfunctions of the symmetric operator $\mathcal{L} = -\Delta$ in $H_0^1(\Omega)$, cf. [19, Chapter 6.5].

We now fix an $N \in \mathbb{N}$ and define the finite dimensional subspace of $H_0^1(\Omega)$ spanned by the first N members of the sequence $\{w_k\}_{k \in \mathbb{N}}$;

$$\mathbb{V}_N := \operatorname{span}\{w_1, \dots, w_N\} \subset H_0^1(\Omega).$$

We aim for constructing a function $u_N : [0, T] \to \mathbb{V}_N$ which approximates $u : [0, T] \to H_0^1(\Omega)$ in space. To this end, u_N shall be of the form

$$u_N(t) := \sum_{k=1}^{N} \alpha_k(t) w_k \qquad (2.17)$$

with time dependent coefficients $\alpha_k : [0, T] \to \mathbb{R}$, $k = 1, \dots, N$. Note that the whole time dependency of $u_N(t)$ is encoded in the coefficients $\alpha_k(t)$ whereas the space dependency is completely contained in the functions w_k. To establish a connection to problem (2.16), we demand that

$$\langle u_N'(t), v \rangle + \mathcal{B}(u_N(t), v) = \langle f(t), v \rangle \quad \text{for all } v \in \mathbb{V}_N,$$
$$\text{f.a.e } t \in [0, T], \quad \text{(Gal-E)}$$
$$\langle u_N(0), v \rangle = \langle u_0, v \rangle \quad \text{f.a.} v \in \mathbb{V}_N. \quad \text{(Gal-I)}$$

Remark 2.2.1 (Relation of u and u_N)
Equation (Gal-I) adopts the initial condition from (2.16) in the sense that it defines $u_N(0) \in \mathbb{V}_N$ as the L^2-projection of $u_0 \in L^2(\Omega)$ to $\mathbb{V}_N$. Similarly, (Gal-E) transfers the differential equation from (2.16) to the finite dimensional space $\mathbb{V}_N$ by considering (2.16a) only on $\mathbb{V}_N$. We understand this in an almost everywhere sense with respect to time.

Next, we show that such a function u_N exists for any $N \in \mathbb{N}$. To that end, we point out that the coefficients $\alpha_k(t)$ are the solutions of a system of ordinary differential equations (ODE), which admits a unique solution according to ODE theory. However, standard ODE theory assumes a continuous right hand side which cannot be provided in the considered case. More general results concerning existence and uniqueness of a solution of an ODE can be found in [5]. For completeness, we present a proof tailored to the considered problem.

Theorem 2.2.2 (Existence and uniqueness of a Galerkin approximation)
For any $N \in \mathbb{N}$ there exists a uniquely defined function $u_N \in H^1(0, T; \mathbb{V}_N)$ of the form (2.17) such that (Gal-E, Gal-I) is satisfied.

Proof. Instead of considering (Gal-E, Gal-I) for all $v \in \mathbb{V}_N$, we may retreat to the basis functions $\{w_k\}_{k=1,\dots,N}$ of $\mathbb{V}_N$. From the structure (2.17) of the function u_N as well as the fact that $\{w_k\}_{k \in \mathbb{N}}$ is an orthonormal basis of $L^2(\Omega)$, we see

$$\langle u_N(0), w_k \rangle = \sum_{l=1}^{N} \langle w_l, w_k \rangle \alpha_l(0) = \alpha_k(0)$$

and

$$\langle u_N'(t), w_k \rangle = \alpha_k'(t).$$

Moreover, we have

$$\mathcal{B}(u_N(t), w_k) = \sum_{l=1}^{N} \mathcal{B}(w_l, w_k)\, \alpha_l(t)$$

and conveniently, we define the matrix $\mathbf{B}$ with $\mathbf{B}_{kl} := \mathcal{B}(w_l, w_k)$ for $k, l = 1, \dots, N$. Further, we define the vectors

$$\alpha(t) := (\alpha_k(t))_{k=1,\dots,N}^\mathsf{T},$$
$$\mathbf{f}(t) := (\langle f(t), w_k \rangle)_{k=1,\dots,N}^\mathsf{T},$$
$$\text{and} \quad \alpha^0 := (\langle u_0, w_k \rangle)_{k=1,\dots,N}^\mathsf{T}.$$

We will shortly use this notation to rewrite (Gal-E, Gal-I) equivalently as ordinary differential equation for the vector $\alpha(t)$. Before doing so, we consider the regularity of $\mathbf{f}$ — which will eventually become the right hand side of the ODE — by observing $\mathbf{f} \in L^2(0, T; \mathbb{R}^N)$ since the components of $\mathbf{f}$ are L^2-regular with respect to time:

$$\|\mathbf{f}_k\|_{L^2(0,T)}^2 = \int_0^T |\mathbf{f}_k(t)|^2 \, dt = \int_0^T |\langle f(t), w_k \rangle|^2 \, dt$$

$$\leq \int_0^T \|f(t)\|_{L^2(\Omega)}^2 \|w_k\|_{L^2(\Omega)}^2 \, dt = \int_0^T \|f(t)\|_{L^2(\Omega)}^2 \, dt$$

$$= \|f\|_{L^2(0,T;L^2(\Omega))}^2.$$

With this regularity and observing that $\mathbf{B}$ is constant with respect to time, we reformulate (Gal-E, Gal-I) as: Find $\alpha \in H^1(0, T; \mathbb{R}^N)$ such that

$$\alpha'(t) + \mathbf{B}\alpha(t) = \mathbf{f}(t) \qquad \text{f.a.e. } t \in (0, T), \qquad \text{(GAL-E')}$$

$$\alpha(0) = \alpha^0. \qquad \text{(GAL-I')}$$

Since the right hand side $\mathbf{f}$ is only L^2-regular with respect to time, we may not employ the standard existence and uniqueness result for ODEs. However, we employ standard methods to establish the existence of a unique solution of (GAL-E', GAL-I').

To that end, we first note that $\alpha \in H^1(0, T; \mathbb{R}^N)$ is a solution of (GAL-E', GAL-I') if and only if $\alpha \in C^0([0, T]; \mathbb{R}^N)$ and it satisfies

$$\alpha(t) = \alpha^0 + \int_0^t \mathbf{f}(s) - \mathbf{B}\alpha(s) \, ds \qquad \text{for } t \in [0, T]. \qquad (2.18)$$

Let $\delta > 0$. We consider the Banach space $\mathbb{X} := C^0([0, T]; \mathbb{R}^N)$ equipped with the norm

$$\|\mathbf{x}\|_\delta := \sup_{t \in [0,T]} e^{-\delta t} |\mathbf{x}(t)|,$$

where $|\cdot|$ denotes the Euclidean norm on $\mathbb{R}^N$. We assume, that the map

$$\mathcal{K} : \mathbb{X} \to \mathbb{X}$$

$$\mathcal{K}(\mathbf{x})(t) := \alpha^0 + \int_0^t \mathbf{f}(s) - \mathbf{B}\mathbf{x}(s)\, ds$$

is a contraction with respect to $\|\cdot\|_\delta$ for some $\delta > 0$. We then deduce by the Banach Fixed Point Theorem that there is a unique $\alpha \in \mathbb{X}$ such that

$$\mathcal{K}(\alpha) = \alpha,$$

or equivalently

$$\alpha(t) = \alpha^0 + \int_0^t \mathbf{f}(s) - \mathbf{B}\alpha(s)\, ds \qquad \text{for } t \in [0, T].$$

Recalling equation (2.18), this implies that α is the unique solution of (GAL-E', GAL-I').

It remains to show that there is a $\delta > 0$ such that $\mathcal{K}$ is contracting. For this purpose, we consider arbitrary $\mathbf{x}, \mathbf{y} \in \mathbb{X}$ and aim at estimating the term $\|\mathcal{K}(\mathbf{x})(t) - \mathcal{K}(\mathbf{y})(t)\|_\delta$. We start by estimating

$$
|\mathcal{K}(\mathbf{x})(t) - \mathcal{K}(\mathbf{y})(t)| = \left| \int_0^t \mathbf{B}\, (\mathbf{y} - \mathbf{x})(s)\, ds \right| \le |\mathbf{B}| \int_0^t |(\mathbf{y} - \mathbf{x})(s)|\, ds
$$

$$
= |\mathbf{B}| \int_0^t e^{\delta s} e^{-\delta s} \, |(\mathbf{y} - \mathbf{x})(s)|\, ds
$$

$$
\le |\mathbf{B}| \, \|\mathbf{y} - \mathbf{x}\|_\delta \int_0^t e^{\delta s}\, ds,
$$

where $|\mathbf{B}|$ denotes the matrix norm of $\mathbf{B}$ that is induced by $|\cdot|$. Hence, for the term involved in $\|\mathcal{K}(\mathbf{x})(t) - \mathcal{K}(\mathbf{y})(t)\|_\delta$ we have

$$
e^{-\delta t} |\mathcal{K}(\mathbf{x})(t) - \mathcal{K}(\mathbf{y})(t)| \le |\mathbf{B}| \, \|\mathbf{y} - \mathbf{x}\|_\delta \int_0^t e^{\delta(s - t)}\, ds
$$

$$
= |\mathbf{B}| \frac{1 - e^{-\delta t}}{\delta} \, \|\mathbf{y} - \mathbf{x}\|_\delta \le |\mathbf{B}| \frac{1}{\delta} \, \|\mathbf{y} - \mathbf{x}\|_\delta\,.
$$

Choosing $\delta := 2\,|\mathbf{B}|$ and employing the definition of $\|\cdot\|_\delta$, we finally derive

$$\|\mathcal{K}(\mathbf{x})(t) - \mathcal{K}(\mathbf{y})(t)\|_\delta = \sup_{t \in [0,T]} e^{-\delta t} |\mathcal{K}(\mathbf{x})(t) - \mathcal{K}(\mathbf{y})(t)|$$

$$\leq \frac{1}{2} \|\mathbf{y} - \mathbf{x}\|_\delta \,,$$

which implies that $\mathcal{K} : \mathbb{X} \to \mathbb{X}$ indeed is a contraction with respect to $\|\cdot\|_\delta$. $\qquad \square$

Now that we established the existence of a unique approximate solution in any given subspace $\mathbb{V}_N \subset H_0^1(\Omega)$, we aim for passing to the limit $N \to \infty$. We will show that a subsequence of the sequence of approximate solutions $\{u_N\}_{N \in \mathbb{N}}$ converges to a weak solution of (2.1). In the process, we will make use of a uniform energy estimate. Before presenting this estimate, we state the following lemma.

Lemma 2.2.3
Assume $v \in H_0^1(\Omega)$ *and let* $v_N := \sum_{k=1}^{N} \langle v, w_k \rangle w_k$ *for the basis* $\{w_k\}_{k \in \mathbb{N}} \subset H_0^1(\Omega)$ *defined above. Then* v_N *satisfies the estimate*

$$\|v_N\|_\Omega \leq \frac{\mathcal{C}_+}{\mathcal{C}_-} \|v\|_\Omega$$

with constants $\mathcal{C}_+$ *and* $\mathcal{C}_-$ *from Lemma 2.1.3.*

Proof. Provided the estimate $\|v_N\|_{H_0^1(\Omega)} \leq \|v\|_{H_0^1(\Omega)}$ holds for the the standard norm on $H_0^1(\Omega)$, we may employ the equivalence to the energy norm $\|\cdot\|_\Omega$ to deduce

$$\|v_N\|_\Omega \leq \mathcal{C}_+ \|v_N\|_{H_0^1(\Omega)} \leq \mathcal{C}_+ \|v\|_{H_0^1(\Omega)} \leq \frac{\mathcal{C}_+}{\mathcal{C}_-} \|v\|_\Omega$$

with constants $\mathcal{C}_+$ and $\mathcal{C}_-$ from Lemma 2.1.3. In the following, we establish the assumed estimate for the standard norm

$$\|v_N\|_{H_0^1(\Omega)} \leq \|v\|_{H_0^1(\Omega)}$$

exploiting the $H_0^1(\Omega)$-orthogonality of the basis $\{w_k\}_{k \in \mathbb{N}}$. We begin with expressing v in terms of the basis $\{w_k\}_{k \in \mathbb{N}} \subset H_0^1(\Omega)$ as

$$v = \lim_{N \to \infty} \sum_{k=1}^{N} \langle v, w_k \rangle w_k.$$

From this representation and exploiting the $H_0^1(\Omega)$-orthogonality of $\{w_k\}_{k \in \mathbb{N}}$ we get

$$
\langle v \,, w_l \rangle_{H_0^1(\Omega)} = \lim_{N \to \infty} \sum_{k=1}^{N} \langle v \,, w_k \rangle \langle w_k \,, w_l \rangle_{H_0^1(\Omega)}
$$

$$
= \langle v \,, w_l \rangle \, \|w_l\|_{H_0^1(\Omega)}^2 . \tag{2.19}
$$

On the other hand, we express

$$
\|v_N\|_{H_0^1(\Omega)}^2 = \langle v_N \,, v_N \rangle_{H_0^1(\Omega)} = \sum_{k,l=1}^{N} \langle \langle v \,, w_k \rangle w_k \,, \langle v \,, w_l \rangle w_l \rangle_{H_0^1(\Omega)}
$$

$$
= \sum_{k,l=1}^{N} \langle v \,, w_k \rangle \langle v \,, w_l \rangle \langle w_k \,, w_l \rangle_{H_0^1(\Omega)} = \sum_{l=1}^{N} \langle v \,, w_l \rangle^2 \, \|w_l\|_{H_0^1(\Omega)}^2
$$

and substituting the term $\langle v \,, w_l \rangle \|w_l\|_{H_0^1(\Omega)}^2$ in this equality via (2.19) we deduce

$$
\|v_N\|_{H_0^1(\Omega)}^2 = \sum_{l=1}^{N} \langle v \,, w_l \rangle \langle v \,, w_l \rangle_{H_0^1(\Omega)} = \langle v \,, v_N \rangle_{H_0^1(\Omega)}
$$

$$
\leq \|v\|_{H_0^1(\Omega)} \, \|v_N\|_{H_0^1(\Omega)}
$$

and hence obtain $\|v_N\|_{H_0^1(\Omega)} \leq \|v\|_{H_0^1(\Omega)}$ as desired. $\qquad \square$

Theorem 2.2.4 (Uniform estimate for Galerkin solutions)
Let $u_N \in H^1(0,T; \mathbb{V}_N)$ be the unique solution of (Gal-E,Gal-I). Then there exists a constant C which is independent of N (and T) such that

$$
\|u_N\|_{C(0,T;L^2(\Omega))} + \|u_N\|_{\mathbb{W}(0,T)} \leq C \Big(\|f\|_{L^2(0,T;L^2(\Omega))}
$$

$$
+ \|u_0\|_{L^2(\Omega)} \Big). \tag{2.20}
$$

Proof. Choosing $w := u_N(t) \in \mathbb{V}_N$ in (Gal-E) we find

$$
\langle u_N'(t) \,, u_N(t) \rangle + \mathcal{B}(u_N(t), u_N(t)) = \langle f(t) \,, u_N(t) \rangle
$$

for almost every $t \in [0, T]$. Since $\langle u_N'(t), u_N(t) \rangle = \frac{1}{2} \frac{d}{dt} \|u_N(t)\|^2$ and employing Cauchy-Schwarz and Young's inequality, we derive

$$\frac{1}{2} \frac{d}{dt} \|u_N(t)\|^2 + \|u_N(t)\|_\Omega^2 = \langle f(t), u_N(t) \rangle$$

$$\leq \frac{1}{2\delta} \|f(t)\|^2 + \frac{\delta}{2} \|u_N(t)\|^2.$$

Choosing $\delta := 1/\mathcal{C}_\mathcal{P}^2$ and estimating $\|u_N(t)\|^2 \leq \mathcal{C}_\mathcal{P}^2 \|u_N(t)\|_\Omega^2$ we find

$$\frac{d}{dt} \|u_N(t)\|^2 + \|u_N(t)\|_\Omega^2 \leq \mathcal{C}_\mathcal{P}^2 \|f(t)\|^2, \tag{2.21}$$

which particularly implies

$$\frac{d}{dt} \|u_N(t)\|^2 \leq \mathcal{C}_\mathcal{P}^2 \|f(t)\|^2.$$

Integrating this with respect to time, we get for $t \in [0, T]$ the pointwise estimate

$$\|u_N(t)\|^2 \leq \|u_N(0)\|^2 + \mathcal{C}_\mathcal{P}^2 \int_0^t \|f(s)\|^2 \, ds. \tag{2.22}$$

Since $u_N(0)$ is the L^2-projection of u_0 onto $\mathbb{V}_N$, we see that $\|u_N(0)\| \leq \|u_0\|$ and taking the maximum over all $t \in (0, T)$ on both sides of (2.22) we obtain

$$\|u_N(t)\|_{C(0,T;L^2(\Omega))}^2 \leq \|u_0\|^2 + \mathcal{C}_\mathcal{P}^2 \|f\|_{L^2(0,T;L^2(\Omega))}^2, \tag{2.23}$$

which is the desired estimate of the first term in (2.20).

In order to derive the remaining estimates, we first integrate inequality (2.21) from 0 to T to particularly find

$$\int_0^T \|u_N(s)\|_\Omega^2 \, ds \leq \|u_N(0)\|^2 + \mathcal{C}_\mathcal{P}^2 \|f\|_{L^2(0,T;L^2(\Omega))}^2$$

$$\leq \|u_0\|^2 + \mathcal{C}_\mathcal{P}^2 \|f\|_{L^2(0,T;L^2(\Omega))}^2. \tag{2.24}$$

To finish the proof, we need to estimate the remaining term $\int_0^T \|u_N(s)\|_*^2 \, ds$ involved in $\|u_N\|_{\mathbb{W}(0,T)}$. For estimating the appearing dual norm, we choose an arbitrary $v \in H_0^1(\Omega)$. Since we may only use test functions from $\mathbb{V}_N$ in (Gal-E), we split v into its L^2-projection $v_N := \sum_{k=1}^{N} \langle v, w_k \rangle w_k \in \mathbb{V}_N$ into $\mathbb{V}_N$ and its orthogonal complement $v_\perp \in H_0^1(\Omega) \setminus \mathbb{V}_N$:

$$v =: v_N + v_\perp \quad \text{with } v_N \in \mathbb{V}_N \text{ and } \langle v_\perp, w_k \rangle = 0 \text{ for } k = 1, \dots, N.$$

From this orthogonality we obtain

$$\langle u_N'(t), v \rangle = \sum_{k=1}^{N} \alpha_k'(t) \langle w_k, v_N + v_\perp \rangle$$

$$= \sum_{k=1}^{N} \alpha_k'(t) \langle w_k, v_N \rangle = \langle u_N'(t), v_N \rangle$$

and applying (Gal-E) we further find

$$= \langle f(t), v_N \rangle - \mathcal{B}(u_N(t), v_N).$$

We estimate these two terms using Cauchy-Schwarz and Poincaré-Friedrichs inequality as well as the continuity of $\mathcal{B}(\cdot, \cdot)$ (cf. Lemma 2.1.3) to obtain

$$|\langle u_N'(t), v \rangle| \leq \|f(t)\| \|v_N\| + \mathcal{C}_\mathcal{B} \|u_N(t)\|_\Omega \|v_N\|_\Omega$$

$$\leq \left(\mathcal{C}_\mathcal{P} \|f(t)\| + \mathcal{C}_\mathcal{B} \|u_N(t)\|_\Omega \right) \|v_N\|_\Omega.$$

Thanks to the representation of the dual pairing between $H^{-1}(\Omega)$ and $H_0^1(\Omega)$ via the L^2-scalar product, and using Lemma 2.2.3 to estimate $\|v_N\|_\Omega$, we get

$$\left| \langle u_N'(t), v \rangle_{H^{-1}(\Omega) \times H_0^1(\Omega)} \right| = |\langle u_N'(t), v \rangle|$$

$$\leq \frac{\mathcal{C}_+}{\mathcal{C}_-} \left(\mathcal{C}_\mathcal{P} \|f(t)\| + \mathcal{C}_\mathcal{B} \|u_N(t)\|_\Omega \right) \|v\|_\Omega.$$

We thus have $\|u_N'(t)\|_* \leq \frac{\mathcal{C}_+}{\mathcal{C}_-} \left(\mathcal{C}_\mathcal{P} \|f(t)\| + \mathcal{C}_\mathcal{B} \|u_N(t)\|_\Omega \right)$ and we obtain

$$\int_0^T \|u_N'(s)\|_*^2 \, ds \leq \frac{\mathcal{C}_+^2}{\mathcal{C}_-^2} \int_0^T 2\mathcal{C}_\mathcal{P}^2 \|f(s)\|^2 + 2\mathcal{C}_\mathcal{B}^2 \|u_N(s)\|_\Omega^2 \, ds$$

$$\leq \frac{\mathcal{C}_+^2}{\mathcal{C}_-^2} \left(2\mathcal{C}_\mathcal{B}^2 \|u_0\|^2 + 2\mathcal{C}_\mathcal{P}^2 (1 + \mathcal{C}_\mathcal{B}^2) \|f\|_{L^2(0,T;L^2(\Omega))}^2 \right).$$

where we used (2.24) in the last estimate. $\qquad\qquad\qquad\qquad\qquad$ $\square$

After this preliminary work, we are now in the position to state existence and uniqueness of a weak solution of (2.1). Making use of the uniform estimate derived in Theorem 2.2.4, we consider the Galerkin approximations $u_N \in \mathbb{V}_N$ and pass to the limit $N \to \infty$ to obtain the exact weak solution.

Theorem 2.2.5 (Existence of a solution)
There exists a weak solution $u \in \mathbb{W}(0,T)$ of (2.1) which satisfies the a priori estimate

$$\|u\|_{C(0,T;L^2(\Omega))} + \|u\|_{\mathbb{W}(0,T)} \leq C \left(\|f\|_{L^2(0,T;L^2(\Omega))} + \|u_0\|_{L^2(\Omega)} \right).$$

The constant C is independent of T.

Proof. Thanks to the uniform energy estimate in Theorem 2.2.4, we see that the sequence

a) $\{u_N\}_{N\in\mathbb{N}}$ is *uniformly* bounded in $L^2(0,T;H_0^1(\Omega))$;

b) $\{u_N'\}_{N\in\mathbb{N}}$ is *uniformly* bounded in $L^2(0,T;H^{-1}(\Omega))$.

Since the spaces $L^2(0,T;H_0^1(\Omega))$ and $L^2(0,T;H^{-1}(\Omega))$ both are Hilbert spaces, they are also reflexive Banach spaces in particular. Hence, there exists (cf. [3, Theorem 6.9]) a subsequence $\{u_{N_l}\}_{l\in\mathbb{N}} \subset \{u_N\}_{N\in\mathbb{N}}$ and functions $u \in L^2(0,T;H_0^1(\Omega))$, $v \in L^2(0,T;H^{-1}(\Omega))$, with

$$u_{N_l} \rightharpoonup u \qquad \text{weakly in } L^2(0,T;H_0^1(\Omega)), \qquad (2.25)$$

$$u_{N_l}' \rightharpoonup v \qquad \text{weakly in } L^2(0,T;H^{-1}(\Omega)). \qquad (2.26)$$

As we will not utilize the two original sequences $\{u_N\}_{N\in\mathbb{N}}$ and $\{u_N'\}_{N\in\mathbb{N}}$ anymore, we conveniently denote the foregoing subsequences by $\{u_N\}_{N\in\mathbb{N}}$ and $\{u_N'\}_{N\in\mathbb{N}}$ respectively. One easily verifies that $v = u'$ and thus $u \in \mathbb{W}(0,T)$. As norms are weakly lower semi-continuous (a consequence of the Hahn-Banach Theorem, cf. [69, Chapter IV,6, Theorem 1 and Corollary 2] or [3, 4.17 Folgerung]), the weak convergence (2.25) implies

$$\|u\|_{L^2(0,T;H_0^1(\Omega))} \leq \liminf_{N\to\infty} \|u_N\|_{L^2(0,T;H_0^1(\Omega))}$$

and applying the uniform bound from Theorem 2.2.4, we deduce

$$\leq C \left(\|f\|_{L^2(0,T;L^2(\Omega))} + \|u_0\|_{L^2(\Omega)} \right).$$

Analogously, we conclude from the weak convergence (2.26) and the uniform bound from Theorem 2.2.4

$$\|u'\|_{L^2(0,T;H^{-1}(\Omega))} \leq \liminf_{N\to\infty} \|u'_N\|_{L^2(0,T;H^{-1}(\Omega))}$$
$$\leq C\left(\|f\|_{L^2(0,T;L^2(\Omega))} + \|u_0\|_{L^2(\Omega)}\right).$$

Combining those two estimates, we bound

$$\|u\|_{\mathbb{W}(0,T)} \leq C\left(\|f\|_{L^2(0,T;L^2(\Omega))} + \|u_0\|_{L^2(\Omega)}\right)$$

and the embedding $\mathbb{W}(0,T) \hookrightarrow C(0,T;L^2(\Omega))$ implies further

$$\|u\|_{C(0,T;L^2(\Omega))} + \|u\|_{\mathbb{W}(0,T)} \leq C\left(\|f\|_{L^2(0,T;L^2(\Omega))} + \|u_0\|_{L^2(\Omega)}\right).$$

Having established the desired estimate, it remains to show that in fact $u \in \mathbb{W}(0,T)$ is a weak solution of (2.1). For that purpose, we fix an integer N_0 and choose a function $v \in C^1(0,T;H_0^1(\Omega))$ of the form

$$v(t) = \sum_{k=1}^{N_0} \alpha_k(t) w_k \tag{2.27}$$

where $\alpha_1(t), \ldots, \alpha_{N_0}(t)$ are given smooth functions. We now choose $N \geq N_0$ and employ $v(t) \in \mathbb{V}_N$ as a test function in (Gal-E). Integrating with respect to time, we obtain

$$\int_0^T \langle u'_N(t), v(t)\rangle + \mathcal{B}(u_N(t), v(t))\, dt = \int_0^T \langle f(t), v(t)\rangle\, dt \tag{2.28}$$

and passing to weak limits utilizing (2.25) and (2.26) we find

$$\int_0^T \langle u'(t), v(t)\rangle + \mathcal{B}(u(t), v(t))\, dt = \int_0^T \langle f(t), v(t)\rangle\, dt. \tag{2.29}$$

Since functions of the form (2.27) are dense in the Lebesgue space $L^2(0,T;H_0^1(\Omega))$, equation (2.29) even holds for all test functions $v \in L^2(0,T;H_0^1(\Omega))$. Hence, in particular

$$\langle u'(t), w\rangle + \mathcal{B}(u(t), w) = \langle f(t), w\rangle$$

holds for each $w \in H_0^1(\Omega)$ and almost every $t \in [0,T]$.

It remains to show that also the initial condition $u(0) = u_0$ is satisfied. To that end, we additionally demand $v(T) = 0$ and observe from (2.28) by integration by parts with respect to time

$$\int_0^T -\langle u_N(t), v'(t)\rangle + \mathcal{B}(u_N(t), v(t))\, dt$$

$$= \int_0^T \langle f(t), v(t)\rangle\, dt + \langle u_N(0), v(0)\rangle.$$

From the initial condition in (Gal-I) we have $\langle u_N(0), v(0)\rangle = \langle u_0, v(0)\rangle$ for arbitrary $N \in \mathbb{N}$ and passing to the weak limit using (2.25) and (2.26), we find

$$\int_0^T -\langle u(t), v'(t)\rangle + \mathcal{B}(u(t), v(t))\, dt$$

$$= \int_0^T \langle f(t), v(t)\rangle\, dt + \langle u_0, v(0)\rangle. \tag{2.30}$$

On the other hand, again by integration by parts with respect to time, equation (2.29) yields

$$\int_0^T -\langle u(t), v'(t)\rangle + \mathcal{B}(u(t), v(t))\, dt = \int_0^T \langle f(t), v(t)\rangle\, dt$$
$$+ \langle u(0), v(0)\rangle. \tag{2.31}$$

Since $v(0) \in H_0^1(\Omega)$ is arbitrary, comparing (2.30) and (2.31) we conclude $u(0) = u_0$ in $L^2(\Omega)$. $\qquad\square$

Theorem 2.2.6 (Uniqueness of a solution)
A weak solution $u \in \mathbb{W}(0, T)$ of (2.1) is unique.

Proof. Let $u_1, u_2 \in \mathbb{W}(0, T)$ be weak solutions of (2.1), i.e., u_1 and u_2 satisfy (2.16). Hence, the difference $u := u_1 - u_2$ of the solutions itself is a solution of the homogeneous problem

$$\langle u', v\rangle + \mathcal{B}(u, v) = 0 \qquad \text{for all } v \in H_0^1(\Omega), \text{ f.a.e. } t \in (0, T)$$
$$u(0) = 0.$$

From the a priori estimate in Theorem 2.2.5 for $f = 0$ and $u_0 = 0$ we deduce

$$\|u\|_{C(0,T;L^2(\Omega))} + \|u\|_{\mathbb{W}(0,T)} \leq 0$$

and hence $u = 0$. $\qquad\square$

2.3   Stability estimate

Later in this work, we introduce an adaptive method for computing a discrete approximation to the exact solution u of (2.1). To
this end, we employ a time stepping scheme with an adaptive control of the current time step size. It lies in the nature of any
time stepping method, that such an adaptive step size control employs temporally *local* information of the (approximate) solution.
More precisely, we anticipate that the step size control considers
the quantity $\|U(t + \tau) - U(t)\|_\Omega$ for some $t \in (0, T)$ and time step
size τ, where U denotes the discrete approximation to the exact
solution u. The crucial point however is that this quantity involves
local information with respect to time, which is of the type $L^\infty - H^1$
rather than global information of the type $L^2 - H^1$. Whereas we
already provided a stability estimate for the latter type by partic-
ularly stating a bound on $\|u\|_{L^2(0,T;H^1(\Omega))}$ in Theorem 2.2.5, we
establish a bound for $\|u\|_{L^\infty(0,T;H^1(\Omega))}$ in Theorem 2.3.3 below. In
Chapter 4, we then present a uniform energy estimate for the discrete approximation of the solution u, which is closely related to
this $L^\infty - H^1$ estimate, see Proposition 4.1.3 and Corollary 4.1.5.

Since the proof of Theorem 2.3.3 employs a splitting of the
bilinear form $\mathcal{B}(\cdot, \cdot)$ into its symmetric and non-symmetric parts,
we state preparatory properties of this splitting in the following
lemma.

Lemma 2.3.1 (Symmetric and non-symmetric parts of $\mathcal{B}$)
Let $v, w \in H_0^1(\Omega)$. We define

$$\mathcal{B}_S(v, w) := \frac{1}{2}\big(\mathcal{B}(v, w) + \mathcal{B}(w, v)\big),$$

$$\mathcal{B}_N(v, w) := \frac{1}{2}\big(\mathcal{B}(v, w) - \mathcal{B}(w, v)\big).$$

*Then $\mathcal{B}_S(\cdot, \cdot)$ and $\mathcal{B}_N(\cdot, \cdot)$ decompose $\mathcal{B}(\cdot, \cdot)$ into its symmetric
and non-symmetric part, i. e., $\mathcal{B}_S(v, w) = \mathcal{B}_S(w, v)$, $\mathcal{B}_N(v, w) \neq
\mathcal{B}_N(w, v)$ and $\mathcal{B}(v, w) = \mathcal{B}_S(v, w) + \mathcal{B}_N(v, w)$. Moreover, it holds
$\mathcal{B}(v, v) = \mathcal{B}_S(v, v)$ and*

$$\mathcal{B}_S(g(t), g'(t)) = \frac{1}{2}\frac{d}{dt}\|g(t)\|_\Omega^2 \qquad \textit{for all } g \in H^1(0, T; H_0^1(\Omega)).$$
$$\tag{2.32}$$

The symmetric part satisfies Cauchy-Schwarz inequality

$$|\mathcal{B}_S(v\,,\,w)| \leq \|v\|_\Omega \, \|w\| \qquad (2.33)$$

and the non-symmetric part satisfies

$$|\mathcal{B}_N(v\,,\,w)| \leq C_{\mathcal{B}_N} \, \|v\|_\Omega \, \|w\| \qquad (2.34)$$

with a constant $C_{\mathcal{B}_N}$ only depending on C_- and $\|\mathbf{b}\|_{L^\infty(\Omega)}$.

Proof. Let $v, w \in H_0^1(\Omega)$. We directly see $\mathcal{B}_S(w\,,\,v) = \mathcal{B}_S(w\,,\,v)$ and $\mathcal{B}_N(v\,,\,w) - \mathcal{B}_N(w\,,\,v) = \mathcal{B}(v\,,\,w) - \mathcal{B}(w\,,\,v) \neq 0$. Likewise, we directly obtain $\mathcal{B}_S(v\,,\,w) + \mathcal{B}_N(v\,,\,w) = \mathcal{B}(v\,,\,w)$ and $\mathcal{B}_S(v\,,\,v) = \mathcal{B}(v\,,\,v)$. For proving the remaining claim (2.32), choose an arbitrary $g \in H^1(0, T; H_0^1(\Omega))$. We then have

$$\frac{d}{dt}\|g(t)\|_\Omega^2 = \frac{d}{dt}\mathcal{B}(g(t)\,,\,g(t)) = \frac{d}{dt}\mathcal{B}_S(g(t)\,,\,g(t))$$
$$= \mathcal{B}_S(g'(t)\,,\,g(t)) + \mathcal{B}_S(g(t)\,,\,g'(t)) = 2\mathcal{B}_S(g(t)\,,\,g'(t)).$$

Since $\mathcal{B}_S(\cdot\,,\,\cdot)$ is a symmetric bilinear form which induces the energy norm, Cauchy-Schwarz inequality (2.33) inequality holds. Further, from the definition of $\mathcal{B}_N(\cdot\,,\,\cdot)$ we find as in the derivation of equation (2.6)

$$\mathcal{B}_N(v\,,\,w) = \frac{1}{2}\int_\Omega \mathbf{b} \cdot \nabla v\, w - \mathbf{b} \cdot \nabla w\, v \, dx = \int_\Omega \mathbf{b} \cdot \nabla v\, w \, dx.$$

From this representation of $\mathcal{B}_N(v\,,\,w)$ we estimate by Cauchy-Schwarz inequality and employing the equivalence of the $H_0^1(\Omega)$-norm and the energy norm

$$|\mathcal{B}_N(v\,,\,w)| \leq \|\mathbf{b} \cdot \nabla v\| \, \|w\| \leq \|\mathbf{b}\|_{L^\infty(\Omega)} \, \|\nabla v\| \, \|w\|$$
$$\leq \frac{\|\mathbf{b}\|_{L^\infty(\Omega)}}{C_-} \, \|v\|_\Omega \, \|w\| =: C_{\mathcal{B}_N} \, \|v\|_\Omega \, \|w\|.$$

$\square$

Similar to the Galerkin method employed in the previous section, we will establish the desired $L^\infty - H^1$ stability estimate in Theorem 2.3.3 by first considering the Galerkin solutions before passing to the limit $n \to \infty$ in a second step. To allow for this passing to the limit, we employ the following lemma.

Lemma 2.3.2

Let $\{v_k\}_{k \in \mathbb{N}} \subset L^2(0, T; H_0^1(\Omega))$ be a sequence converging weakly to some $v \in L^2(0, T; H_0^1(\Omega))$. Suppose also that $\operatorname{ess\,sup}_{t \in [0,T]} \|v_k(t)\|_\Omega$ is uniformly bounded, i. e., there exists a constant C_1 such that

$$\operatorname*{ess\,sup}_{t \in [0,T]} \|v_k(t)\|_\Omega \leq C_1 \qquad \text{for all } k \in \mathbb{N}.$$

Then also $\operatorname{ess\,sup}_{t \in [0,T]} \|v(t)\|_\Omega$ is bounded by the same constant, i. e.,

$$\operatorname*{ess\,sup}_{t \in [0,T]} \|v(t)\|_\Omega \leq C_1.$$

Proof. Let $w \in L^2(0, T; H_0^1(\Omega))$. The weak convergence particularly implies for the symmetric bilinear form $\mathcal{B}_S(\cdot, \cdot)$ from Lemma 2.3.1

$$\int_0^T \mathcal{B}_S(v(t), w(t)) \, dt = \lim_{k \to \infty} \int_0^T \mathcal{B}_S(v_k(t), w(t)) \, dt$$

$$\leq \lim_{k \to \infty} \int_0^T \|v_k(t)\|_\Omega \|w(t)\|_\Omega \, dt,$$

where we used Cauchy-Schwarz inequality. Combining this estimate with the uniform bound on $\operatorname{ess\,sup}_{t \in [0,T]} \|v_k(t)\|_\Omega$ for all $k \in \mathbb{N}$, we obtain the upper bound

$$\int_0^T \mathcal{B}_S(v(t), w(t)) \, dt \leq C_1 \int_0^T \|w(t)\|_\Omega \, dt. \qquad (2.35)$$

To establish the bound on $\operatorname{ess\,sup}_{t \in [0,T]} \|v(t)\|_\Omega$, we argue by contradiction and assume

$$\operatorname*{ess\,sup}_{t \in [0,T]} \|v(t)\|_\Omega > C_1.$$

Consequently, there exists a set $\mathcal{A} \subset [0, T]$ with positive measure $|\mathcal{A}| > 0$ such that

$$\|v(t)\|_\Omega > C_1 \qquad \text{for almost all } t \in \mathcal{A}.$$

In this setting, we choose w explicitly as

$$w := \alpha \, v \qquad \text{with} \qquad \alpha(t) := \begin{cases} 1 & \text{for } t \in \mathcal{A} \\ 0 & \text{for } t \notin \mathcal{A} \end{cases} \qquad (2.36)$$

and since $\mathcal{A}$ is measurable, we have $\alpha \in L^\infty(0,T)$ and consequently, $\alpha\, v \in L^2(0,T; H_0^1(\Omega))$ is an admissible choice for w. With this choice, we deduce

$$\int_0^T \mathcal{B}_S(v(t)\,,\,w(t))\,dt = \int_0^T \mathcal{B}_S(v(t)\,,\,\alpha(t)\,v(t))\,dt = \int_{\mathcal{A}} \|v(t)\|_\Omega^2\,dt.$$

Using the assumption $\|v(t)\|_\Omega > C_1$ for almost all $t \in \mathcal{A}$ as well as the definition of w and $\alpha(t)$ from (2.36), we further deduce

$$> C_1 \int_{\mathcal{A}} \|v(t)\|_\Omega\,dt = C_1 \int_0^T \|w(t)\|_\Omega\,dt,$$
$$(2.37)$$

a contradiction to (2.35). Hence, we conclude $\operatorname{ess\,sup}_{t \in [0,T]} \|v(t)\|_\Omega \le C_1$. $\qquad\square$

After these preliminaries, we are in the position to state the following stability estimate.

Theorem 2.3.3 (Stability estimate)
Let $u \in \mathbb{W}(0,T)$ be the weak solution of (2.16) and assume that we have compatible initial data $u_0 \in H_0^1(\Omega)$. Then the following estimate holds:

$$\|u'\|_{L^2(0,T;L^2(\Omega))}^2 + \operatorname*{ess\,sup}_{t \in [0,T]} \|u(t)\|_\Omega^2 \le 2 \Big(2\mathcal{C}_{\mathcal{B}_N}^2 \|u_0\|_{L^2(\Omega)}^2$$

$$+ \frac{\mathcal{C}_+^2}{\mathcal{C}_-^2} \|u_0\|_\Omega^2 + (2 + 2\mathcal{C}_{\mathcal{B}_N}^2 \mathcal{C}_{\mathcal{P}}^2) \|f\|_{L^2(0,T;L^2(\Omega))}^2 \Big).$$

Proof. We choose a fixed N as the dimension of the Galerkin approximation space. From Theorem 2.2.2 we recall $u_N \in H^1(0,T; \mathbb{V}_N)$ and thus, we have $u_N'(t) \in \mathbb{V}_N$ for almost every $t \in (0,T)$. Accordingly, we may choose $u_N'(t)$ as a test function in (Gal-E) to obtain

$$\langle u_N'(t)\,,\,u_N'(t)\rangle + \mathcal{B}(u_N(t)\,,\,u_N'(t)) = \langle f(t)\,,\,u_N'(t)\rangle$$

for a. e. $t \in [0,T]$. Splitting the bilinear form $\mathcal{B}(\cdot\,,\,\cdot)$ into its symmetric and non-symmetric parts in the fashion of Lemma 2.3.1, this equivalently reads

$$\langle u'_N(t)\,,\,u'_N(t)\rangle + \mathcal{B}_S(u_N(t)\,,\,u'_N(t)) = \langle f(t)\,,\,u'_N(t)\rangle$$
$$-\mathcal{B}_N(u_N(t)\,,\,u'_N(t))$$

for a. e. $t \in [0,T]$. We may then employ $\mathcal{B}_S(u_N(t)\,,\,u'_N(t)) = \frac{1}{2}\frac{d}{dt}\,\|u_N(t)\|_\Omega^2$ from Lemma 2.3.1 to get

$$\|u'_N(t)\|^2 + \frac{1}{2}\frac{d}{dt}\,\|u_N(t)\|_\Omega^2 = \langle f(t)\,,\,u'_N(t)\rangle - \mathcal{B}_N(u_N(t)\,,\,u'_N(t))$$

for a. e. $t \in [0,T]$. Integrating with respect to time yields for an arbitrary $t \in (0,T)$

$$\int_0^t \|u'_N(s)\|_{L^2(\Omega)}^2\,ds + \frac{1}{2}\,\|u_N(t)\|_\Omega^2 - \frac{1}{2}\,\|u_N(0)\|_\Omega^2$$

$$= \int_0^t \langle f(s)\,,\,u'_N(s)\rangle - \mathcal{B}_N(u_N(s)\,,\,u'_N(s))\,ds,$$

which can be further estimated using Cauchy-Schwarz and Young's inequalities as well as the continuity (2.34) of the non-symmetric part from Lemma 2.3.1;

$$\leq \|f\|_{L^2(0,t;L^2(\Omega))}\,\|u'_N\|_{L^2(0,t;L^2(\Omega))}$$

$$+ C_{\mathcal{B}_N}\,\|u_N\|_{L^2(0,t;H_0^1(\Omega))}\,\|u'_N\|_{L^2(0,t;L^2(\Omega))}$$

$$\leq \|f\|_{L^2(0,t;L^2(\Omega))}^2 + \frac{1}{4}\,\|u'_N\|_{L^2(0,t;L^2(\Omega))}^2$$

$$+ C_{\mathcal{B}_N}^2\,\|u_N\|_{L^2(0,t;H_0^1(\Omega))}^2 + \frac{1}{4}\,\|u'_N\|_{L^2(0,t;L^2(\Omega))}^2\,.$$

Noticing that the last term on the right hand side is of the same type as the first term on the left hand side, we add $\frac{1}{2}\,\|u_N(0)\|_\Omega^2 - \frac{1}{2}\,\|u'_N\|_{L^2(0,t;L^2(\Omega))}^2$ on both sides of the foregoing estimate. This yields

$$\|u'_N\|_{L^2(0,t;L^2(\Omega))}^2 + \|u_N(t)\|_\Omega^2 \leq \|u_N(0)\|_\Omega^2 + 2\,\|f\|_{L^2(0,t;L^2(\Omega))}^2$$

$$+ 2C_{\mathcal{B}_N}^2\,\|u_N\|_{L^2(0,t;H_0^1(\Omega))}^2$$

$$\leq \|u_N(0)\|_\Omega^2 + 2\,\|f\|_{L^2(0,T;L^2(\Omega))}^2$$

$$+ 2C_{\mathcal{B}_N}^2\,\|u_N\|_{L^2(0,T;H_0^1(\Omega))}^2\,.$$

From Lemma 2.2.3, we further estimate $\|u_N(0)\|_\Omega \leq \mathcal{C}_+/\mathcal{C}_- \|u_0\|_\Omega$ and we use estimate (2.24) from the proof of Theorem 2.2.4 to bound $\|u_N\|^2_{L^2(0,T;H^1_0(\Omega))} \leq \|u_0\|^2_{L^2(\Omega)} + \mathcal{C}_\mathcal{P}^2 \|f\|^2_{L^2(0,T;L^2(\Omega))}$. This implies the uniform estimate

$$\|u'_N\|^2_{L^2(0,t;L^2(\Omega))} + \|u_N(t)\|^2_\Omega \leq 2\mathcal{C}_{\mathcal{B}_N}^2 \|u_0\|^2_{L^2(\Omega)} + \frac{\mathcal{C}_+^2}{\mathcal{C}_-^2} \|u_0\|^2_\Omega$$

$$+(2 + 2\mathcal{C}_{\mathcal{B}_N}^2 \mathcal{C}_\mathcal{P}^2) \|f\|^2_{L^2(0,T;L^2(\Omega))} =: \text{rhs},$$

where we conveniently abbreviate the right hand side by rhs to improve readability in the following. Considering that the choice of $t \in [0, T]$ is arbitrary, we find

$$\|u'_N\|^2_{L^2(0,T;L^2(\Omega))} + \sup_{t\in[0,T]} \|u_N(t)\|^2_\Omega \leq \text{rhs}. \qquad (2.38)$$

This particularly states a uniform bound for $\|u'_N\|_{L^2(0,T;L^2(\Omega))}$. Since $L^2(0, T; L^2(\Omega))$ is a Hilbert space and hence especially a reflexive Banach space, we deduce as in the proof of Theorem 2.2.5 that there exists a subsequence $\{u'_{N_l}\}_{l\in\mathbb{N}} \subset \{u'_N\}_{N\in\mathbb{N}}$ such that $\{u'_{N_l}\}_{l\in\mathbb{N}}$ converges weakly to u' in $L^2(0, T; L^2(\Omega))$. Analogously to the proof of Theorem 2.2.5, we conclude by the weak lower semicontinuity of the norm and the uniform estimate (2.38)

$$\|u'\|^2_{L^2(0,T;L^2(\Omega))} \leq \text{rhs}. \qquad (2.39)$$

With this estimate at hand, it remains to establish the desired bound for $\operatorname{ess\,sup}_{t\in[0,T]} \|u(t)\|^2_\Omega$. Note that we cannot employ an argument similar to the one just used, since (2.38) gives a uniform bound in $L^\infty(0, T; H^1_0(\Omega))$ which is not reflexive and hence does not imply the existence of a weakly convergent subsequence. Instead, we recall from Theorem 2.2.4 that the sequence $\{u_N\}_{N\in\mathbb{N}}$ is uniformly bounded in the Hilbert space $L^2(0, T; H^1_0(\Omega))$. Hence, as in the proof of Theorem 2.2.5 we deduce that there exists a subsequence $\{u_{N_l}\}_{l\in\mathbb{N}} \subset \{u_N\}_{N\in\mathbb{N}}$ which converges weakly to u in $L^2(0, T; H^1_0(\Omega))$. Combining this weak convergence with the uniform bound

$$\operatorname{ess\,sup}_{t\in[0,T]} \|u_{N_l}(t)\|^2_\Omega \leq \sup_{t\in[0,T]} \|u_{N_l}(t)\|^2_\Omega \leq \text{rhs}$$

derived from (2.38), we employ Lemma 2.3.2 to find

$$\operatorname*{ess\,sup}_{t \in [0,T]} \|u(t)\|_\Omega^2 \le \text{rhs.} \qquad (2.40)$$

Combining estimates (2.39) and (2.40) proves the claim. □

Remark 2.3.4 (Constants in continuous and discrete stability estimate)
As indicated, the stability estimate of the foregoing Theorem 2.3.3 is closely related to the uniform energy estimate of Proposition 4.1.3. Comparing those two estimates we realize that also the involved constants are quite similar as both estimates employ the same constants related to the L^2-norm of the initial value as well as the $L^2 - L^2$-norm of the right hand side. However, we notice that the continuous estimate of Theorem 2.3.3 includes an additional factor 2 for the whole estimate which is due to the fact that the bounds on $\|u'\|^2_{L^2(0,T;L^2(\Omega))}$ and $\operatorname{ess\,sup}_{t \in [0,T]} \|u(t)\|_\Omega^2$ are established individually. Moreover, we point out that the energy norm of the initial value contains an additional factor $\mathcal{C}_+^2/\mathcal{C}_-^2$ in the foregoing theorem. This factor, however, is not related to the nature of the energy estimate but rather arises from estimating $\|u_N(0)\|_\Omega^2$ via Lemma 2.2.3. We mention that also a constant-free version of Lemma 2.2.3, i. e., an estimate of the form $\|u_N(0)\|_\Omega^2 \le \|u_0\|_\Omega^2$ can be derived by employing different finite dimensional spaces in the Galerkin method which are tailored to the employed differential operator $\mathcal{L}$.

Chapter 3

Discretization of parabolic PDEs

This chapter is devoted to discretizing the parabolic partial differential equation (2.1) introduced and analyzed in Chapter 2. To this end, we use finite elements for spatial discretization and the implicit Euler method for discretization with respect to time. Before stating this discretization precisely in Section 3.2, we introduce finite element spaces and their underlying triangulations in Section 3.1. As we will see in Section 3.2, the employed full space time discretization splits the time dependent problem (2.1) into a sequence of stationary problems. Owing to that, we include a brief digression into adaptive methods for elliptic PDEs in Section 3.3. Returning to parabolic problems, we conclude this chapter by presenting a posteriori error estimates in Section 3.4.

3.1 Triangulations and finite element spaces

We first define the concepts conforming triangulations and finite element spaces. The definitions and concepts are taken from [15, 9, 11, 57].

Definition 3.1.1 (Simplex)
For $n \in \mathbb{N}_0$ with $0 \leq n \leq d$, consider a set of $n + 1$ vectors $\{a_0, \ldots, a_n\} \subset \mathbb{R}^d$. We assume that the n vectors $a_1 - a_0, \ldots, a_n - a_0$ are linear independent in $\mathbb{R}^d$.

a) The set

$$E = \text{conv hull}\,\{a_0, \ldots, a_n\} = \left\{ \sum_{i=0}^{n} \lambda_i a_i \mid \lambda_i \geq 0 \text{ and } \sum_{i=0}^{n} \lambda_i = 1 \right\},$$

is known as the *n-simplex* spanned by $a_0, \ldots, a_n$. The coefficients λ_i describing a point $x \in E$ are unique and known as the *barycentric coordinates* of x in the simplex E.

b) For $k \in \mathbb{N}_0$ with $0 \leq k \leq n$, let E' be a k-simplex spanned by $a'_0, \ldots, a'_k \in \{a_0, \ldots, a_n\}$. Then E' is called a *k-sub-simplex* of E. The 0-sub-simplices are the *vertices* of E and the 1-sub-simplices are called the *sides* of E.

c) Let $e_1, \ldots, e_d$ denote the canonical basis of $\mathbb{R}^d$. The d-simplex

$$\hat{E} := \text{conv hull}\,\{0, e_1, \ldots, e_d\}$$

is called the *reference simplex* in $\mathbb{R}^d$.

d) For an n-simplex E we define

the mesh size $\qquad h_E := |E|^{1/n}$;

the diameter $\qquad \overline{h}_E := \sup\{|x - y| \mid x, y \in E\}$;

the inball diameter $\quad \underline{h}_E := \sup\{2r \mid B_r \subset E \text{ is a sphere of radius } r\}$;

the shape coefficient $\quad \sigma_E := \dfrac{\overline{h}_E}{\underline{h}_E}$.

For any d-simplex $E := \text{conv hull}\,\{a_0, \ldots, a_d\}$ in $\mathbb{R}^d$, there exists a bijective affine mapping $F_E : \hat{E} \to E$ represented via

$$F_E \hat{x} := A_E \hat{x} + a_0$$

with

$$A_E := \begin{pmatrix} \vdots & & \vdots \\ a_1 - a_0 & \cdots & a_d - a_0 \\ \vdots & & \vdots \end{pmatrix} \in \mathbb{R}^{d \times d}.$$

It holds

$$\|A_E\| \leq \frac{\overline{h}_E}{\underline{h}_{\hat{E}}}, \qquad \|A_E^{-1}\| \leq \frac{\overline{h}_{\hat{E}}}{\underline{h}_E}, \qquad |\det A_E| = \frac{|E|}{|\hat{E}|},$$

where $\|\cdot\|$ denotes the spectral norm on $\mathbb{R}^{d \times d}$, see e.g., [11, 15, 40, 57, 9].

Definition 3.1.2 (Triangulation)
Let $\mathcal{G}$ be a finite set of d-simplices in $\mathbb{R}^d$.

a) The set $\mathcal{G}$ is called a *triangulation* of Ω if

$$\overline{\Omega} = \bigcup_{E \in \mathcal{G}} E \qquad \text{and} \qquad |\Omega| = \sum_{E \in \mathcal{G}} |E|.$$

b) A triangulation $\mathcal{G}$ is called *conforming* if the intersection $E_1 \cap E_2$ of any two elements $E_1, E_2 \in \mathcal{G}$ is either empty or a complete k-sub-simplex of both E_1 and E_2, with $0 \leq k \leq d$.

c) A sequence $\{\mathcal{G}_i\}_{i \in \mathbb{N}}$ of conforming triangulations of Ω is called *shape regular* if the shape coefficient $\sigma_{\mathcal{G}} := \max_{E \in \mathcal{G}} \sigma_E$ is uniformly bounded for all involved triangulations, i.e.,

$$\sup_{i \in \mathbb{N}} \sigma_{\mathcal{G}_i} < C$$

for a constant $C < \infty$ which is independent of i.

d) For a simplex $E \in \mathcal{G}$, we denote the set of its vertices by $\mathcal{V}_E$ and the set of its sides by $\mathcal{S}_E$. Conveniently, we abbreviate the set of all vertices resp. sides in a triangulation $\mathcal{G}$ as

$$\mathcal{V}_{\mathcal{G}} := \bigcup_{E \in \mathcal{G}} \mathcal{V}_E \qquad \text{and} \qquad \mathcal{S}_{\mathcal{G}} := \bigcup_{E \in \mathcal{G}} \mathcal{S}_E.$$

Moreover, we define the set of interior vertices $\overset{\circ}{\mathcal{V}}_{\mathcal{G}} := \{v \in \mathcal{V}_{\mathcal{G}} \mid v \in \Omega\}$ as well as the set of interior sides $\overset{\circ}{\mathcal{S}}_{\mathcal{G}} := \{S \in \mathcal{S}_{\mathcal{G}} \mid S \cap \Omega \neq \emptyset\}$. The *skeleton* of $\mathcal{G}$ is defined as

$$\Gamma_{\mathcal{G}} := \bigcup_{S \in \mathcal{S}_{\mathcal{G}}} \{x \mid x \in S\} \subset \overline{\Omega}.$$

Remark 3.1.3 (Finite element nomenclature)
Owing to usual finite element nomenclature, we mainly refer to a simplex $E \in \mathcal{G}$ as an *element*. Further, we also refer to a conforming triangulation $\mathcal{G}$ as a *mesh* or *grid* and to a vertex $v \in \mathcal{V}_{\mathcal{G}}$ as a *node*.

Associated to vertices, sides, and elements, we define the following patches, an illustration can be found in Figure 3.1.

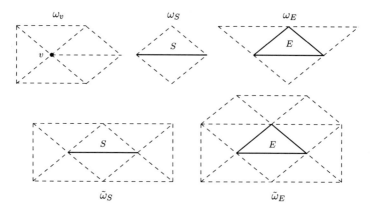

Figure 3.1: Illustration of patches in 2d associated to an interior node, side and element (top row) as well as enlarged patches associated to an interior side and element (bottom row).

Definition 3.1.4 (Patches)
For a triangulation $\mathcal{G}$, we define

a) the patch of all elements associated with a vertex $v \in \mathcal{V}_{\mathcal{G}}$

$$\omega_v := \{E \in \mathcal{G} \mid v \in \mathcal{V}_E\};$$

b) the patch of all elements associated with a side $S \in \mathcal{S}_{\mathcal{G}}$

$$\omega_S := \{E \in \mathcal{G} \mid S \in \mathcal{S}_E\};$$

c) the patch of all elements associated with an element $E \in \mathcal{G}$

$$\omega_E := \{E' \in \mathcal{G} \mid \mathcal{S}_{E'} \cap \mathcal{S}_E \neq \emptyset\}.$$

Further, we define the *enlarged* patches associated with a side or an element, which relax the condition that a whole side is shared and only require non-empty intersections in the following sense. We define

d) the enlarged patch of all elements associated with a side $S \in \mathcal{S}_{\mathcal{G}}$

$$\tilde{\omega}_S := \{E \in \mathcal{G} \mid S \cap E \neq \emptyset\};$$

e) the enlarged patch of all elements associated with an element $E \in \mathcal{G}$

$$\tilde{\omega}_E := \{E' \in \mathcal{G} \mid E' \cap E \neq \emptyset\}.$$

3.1.1 Refinement and coarsening

A triangulation of the domain Ω will be the foundation of the finite element spaces introduced shortly in Section 3.1.2. Roughly speaking, the employed finite element space consists of continuous functions $V : \Omega \to \mathbb{R}$ which are piecewise affine with respect to the used triangulation $\mathcal{G}$. Starting in Section 3.2, we aim at approximating the weak solution of (2.1) in such a finite element space throughout this work. As intuitively the size of the individual triangles is closely related to the quality of the approximation, it is of utmost importance to have at hand routines for locally refining and coarsening a triangulation. In the following, we briefly review a widespread refinement technique and introduce an abstract module implementing the same. For a detailed review, we refer to [40, 57] and the references therein.

Refining a single simplex

Since any triangulation in $\mathbb{R}^d$ consists of d-simplices, we need to state how a single simplex is refined. In this work, we use refinement by *bisection*, i.e., a given simplex E is split into two commensurate sub-simplices, cf. [6, 30, 4, 26, 32]. In order to keep the shape coefficient σ bounded, this subdivision is *not* arbitrary,

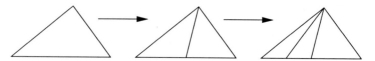

Figure 3.2: Successive refinement leading to degenerated simplices with unbounded shape coefficient σ.

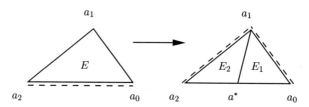

Figure 3.3: Bisection of a two-dimensional simplex E. The dashed lines mark the refinement edges.

see Figure 3.2, but rather is subject to certain rules. In the following, we briefly show how a simplex in 2d can be refined such that shape regularity is preserved. For details and especially higher dimensional refinements, we refer to [62, 40].

We identify a 2-simplex E with the set of its *ordered* vertices, $E = \{a_0, a_1, a_2\}$. In order to bisect E into two simplices E_1 and E_2, we first specify the edge $\overline{a_0 a_2}$ to be the *refinement edge*, meaning that the vertex to be newly created is given by the midpoint $a^* := (a_0 + a_2)/2$ of this edge. The two new simplices E_1 and E_2 share this new vertex a^* and the whole side defined by the nodes $\{a^*, a_1\}$. They are given by

$$E_1 = \{a_0, a^*, a_1\} \qquad \text{and} \qquad E_2 = \{a_2, a^*, a_1\}.$$

Particularly, the refinement edges of both children E_1 and E_2 lie *opposed* to the newly created vertex a^*, which suggests that shape regularity is preserved, compare Figure 3.3.

Conveniently, we assume we have at hand a function BISCET implementing shape regular bisection in any dimension: Given any simplex E,

$$\{E_1, E_2\} = \mathsf{BISCET}(E)$$

outputs two simplices E_1 and E_2 such that shape regularity is preserved. For details on such a method, we refer to [62]. As the input of BISCET is an arbitrary simplex, recurrent application is possible. Starting from an initial simplex E_0, this iterative process may conveniently be represented as an (infinite) binary tree $\mathbb{F}(E_0)$. In this tree, the initial simplex E_0 constitutes the root while the nodes $E \in \mathbb{F}(E_0)$ correspond to the simplices created by repeated application of BISCET. For an illustration see Figure 3.4.

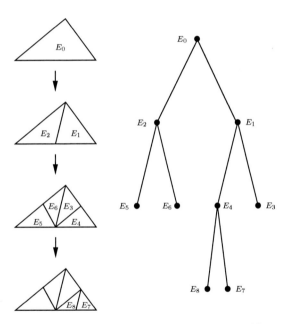

Figure 3.4: Refinements of a simplex E and associated binary tree $\mathbb{F}(E_0)$.

Definition 3.1.5 (Generation)
We define the *generation* $g(E)$ of a node $E \in \mathbb{F}(E_0)$ as the number of bisections needed to produce E from E_0. The generation can equivalently be expressed as the number of predecessors in the tree.

Refining a triangulation

Having introduced the refinement of a single simplex, we now turn to the refinement of a whole triangulation. Analogously to the association of the recurrent bisection of a single simplex with a binary tree, we connect a refinement of an initial triangulation $\mathcal{G}_{\text{init}}$ with a *forest* of binary trees. All possible refinements of an initial triangulation $\mathcal{G}_{\text{init}}$ are associated with a *master forest* of binary trees.

Definition 3.1.6 (Forest and refinement)
Let $\mathcal{G}_{\text{init}}$ be a conforming initial triangulation. We then define the associated *master forest* of binary trees as

$$\mathbb{F} = \mathbb{F}(\mathcal{G}_{\text{init}}) := \bigcup_{E_0 \in \mathcal{G}_{\text{init}}} \mathbb{F}(E_0).$$

The *generation* of a node $E \in \mathbb{F}$ with $E \in \mathbb{F}(E_0)$, $E_0 \in \mathcal{G}_{\text{init}}$, is defined as the generation $g(E)$ of E within $\mathbb{F}(E_0)$. A subset $\mathcal{F} \subset \mathbb{F}$ is called *forest* if

a) it contains the initial triangulation, i.e., $\mathcal{G}_{\text{init}} \subset \mathcal{F}$;

b) all nodes of $\mathcal{F} \setminus \mathcal{G}_{\text{init}}$ have a predecessor;

c) all nodes of $\mathcal{F}$ have either two successors or none.

A forest $\mathcal{F}$ is called *finite*, if $\max_{E \in \mathcal{F}} g(E) < \infty$. A node $E \in \mathcal{F}$ is called *root* if it possesses no predecessor (equivalently $g(E) = 0$), it is called *leaf* if it possesses no successor.

With this definition, it is clear that the roots of any forest $\mathcal{F} \subset \mathbb{F}(\mathcal{G}_{\text{init}})$ are the simplices of $\mathcal{G}_{\text{init}}$. Additionally, we may uniquely associate any forest $\mathcal{F} \subset \mathbb{F}(\mathcal{G}_{\text{init}})$ with a triangulation $\mathcal{G} = \mathcal{G}(\mathcal{F})$ of Ω by defining $\mathcal{G}$ to be the set of leaves of $\mathcal{F}$. This motivates the following definition of a partial order of triangulations.

Definition 3.1.7 (Refinements)
Let $\mathcal{G}_{\text{init}}$ be a triangulation of Ω and let $\mathcal{F}, \mathcal{F}_+ \subset \mathbb{F}(\mathcal{G}_{\text{init}})$ be two finite forests with associated triangulations $\mathcal{G}$ and $\mathcal{G}_+$. The triangulation $\mathcal{G}_+$ is called a *refinement* of $\mathcal{G}$ if $\mathcal{F} \subset \mathcal{F}_+$. We denote $\mathcal{G}_+$ being a refinement of $\mathcal{G}$ by $\mathcal{G} \leq \mathcal{G}_+$ or equivalently $\mathcal{G}_+ \geq \mathcal{G}$.

We emphasize that while refining all simplices in $\mathcal{G}$ in order to give rise to $\mathcal{G}_+$ (so-called *uniform* or *global* refinement) certainly implies $\mathcal{G}_+ \geq \mathcal{G}$, this is not a necessary condition. We may rather refine only some simplices leaving the others untouched and observe that this also produces a refinement $\mathcal{G}_+$ of $\mathcal{G}$. This selective refinement is very important in adaptive processes.

Remark 3.1.8 (Partial order)
Apparently, the set inclusion "⊂" defines a partial order on the master forest $\mathbb{F}(\mathcal{G}_{\text{init}})$. In the light of Definition 3.1.7, this structure directly transfers to the relation "is refinement of" between triangulations that originate from the same macro triangulation $\mathcal{G}_{\text{init}}$. The notations "≤" and "≥" intuitively represent this partial order.

Remark 3.1.9 (Conformity)
Note that whereas the definition of a forest implies that the leaf nodes cover Ω and consequently constitute a triangulation, it is not clear if this triangulation is conforming. In fact, conformity cannot be expected in general. Apparently, selective refinement of some simplices of a triangulation may violate conformity, see Figure 3.5. More unexpectedly, also global refinement may lead to non-conforming triangulations, see Figure 3.6. However, in 2d two consecutive global refinements bisect every side $S \in \mathcal{S}_{\mathcal{G}}$ exactly once and the midpoints are the vertices of the grandchildren. Hence, a conforming triangulation is created, see Figure 3.6. Opposed to this quite transparent situation in 2d, higher dimensions do require additional conditions on the initial triangulation $\mathcal{G}_{\text{init}}$ and particularly on the choice of refinement edges to guarantee the existence of a conforming refinement. For an overview, we refer to [40], further details can be found in [6, 30, 4].

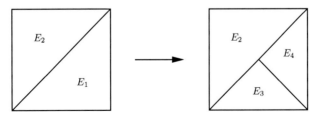

Figure 3.5: Selective refinement (only E_1 is refined) of a triangulation violating conformity.

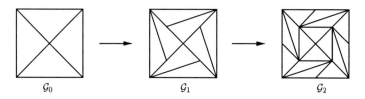

$\mathcal{G}_0$ $\qquad\qquad\qquad\qquad$ $\mathcal{G}_1$ $\qquad\qquad\qquad\qquad$ $\mathcal{G}_2$

Figure 3.6: Uniform refinement of a triangulation $\mathcal{G}_0$ into $\mathcal{G}_1$ violating conformity. Conformity is recovered after subsequent uniform refinement of $\mathcal{G}_1$ into $\mathcal{G}_2$.

With regard to adaptive algorithms, we are especially interested in successively modifying a given conforming initial triangulation $\mathcal{G}_{\text{init}}$. We define the class of conforming refinements of $\mathcal{G}_{\text{init}}$ as

$$\mathbb{G} := \{\mathcal{G} = \mathcal{G}(\mathcal{F}) \mid \mathcal{F} \subset \mathbb{F}(\mathcal{G}_{\text{init}}) \text{ is finite and } \mathcal{G}(\mathcal{F}) \text{ is conforming}\}.$$

In the 2d case, we see from Remark 3.1.9 that $\mathbb{G}$ is non-empty, which we also assume for higher dimensions. For conditions in higher dimensions we refer to [6, 30, 4].

In adaptive applications, we are particularly interested in refining only a certain subset of a triangulation. For this purpose, we collect the elements to be refined in the set of *marked* elements $\mathcal{M} \subset \mathcal{G}$. It is also convenient to specify how many times a given marked element shall be refined. To that end, we associate with each marked element $E \in \mathcal{M}$ an *element marker* $m(E) \in \mathbb{N}$ and demand that any element $E \in \mathcal{M}$ is refined at least $m(E)$ times. With this in mind, the complete information about a desired refinement of a mesh $\mathcal{G}$ is given by the set

$$\mathcal{M}^{\text{m}} := \{(E, m(E)) \mid E \in \mathcal{M}\} \subset \mathcal{G} \times \mathbb{N}.$$

For a conforming triangulation $\mathcal{G}$ and a set $\mathcal{M}^{\text{m}} \subset \mathcal{G} \times \mathbb{N}$ of marked elements and associated element markers, we then assume we have at hand a function REFINE, which produces a conforming triangulation $\mathcal{G}_+ \geq \mathcal{G}$ where any element $E \in \mathcal{M}$ was bisected at least $m(E)$ times;

$$\mathcal{G}_+ = \text{REFINE}(\mathcal{M}^{\text{m}}, \mathcal{G}).$$

Remark 3.1.10 (Refinement algorithms)
In principle, there are two different ways of realizing a function RE-

FINE with the above properties; iterative refinement and recursive refinement.

The idea behind the iterative version is to first exclusively refine all marked elements and re-establish conformity in a second step. This is accomplished by identifying all elements infringing conformity by containing a hanging node and marking them for refinement before entering the next iterative round.

The recursive version on the other hand only refines a single simplex if this atomic refinement does *not* violate conformity, i. e., no hanging node is produced. If the refinement of a marked element however does invoke a hanging node, the algorithm first tries to refine the appropriate neighbor such that the potential hanging node becomes a regular one. This process is carried out recursively.

For a thorough review of those procedures, we refer to [40, Chapter 4]. Details on implementational aspects can be found in [57, Section 3.4].

Coarsening a triangulation

Intuitively, we understand a coarsening of a triangulation as the opposite of a refinement. More precisely, building upon the definition of a refinement (cf. Definition 3.1.7), we define:

Definition 3.1.11 (Coarsening)
Let $\mathcal{G}_{\text{init}}$ be a triangulation of Ω and let $\mathcal{F}, \mathcal{F}_- \subset \mathbb{F}(\mathcal{G}_{\text{init}})$ be two finite forests with associated triangulations $\mathcal{G}$ and $\mathcal{G}_-$. The triangulation $\mathcal{G}_-$ is called a *coarsening* of $\mathcal{G}$ if $\mathcal{F}_- \subset \mathcal{F}$. Equivalently, $\mathcal{G}_-$ is a coarsening of $\mathcal{G}$ iff $\mathcal{G}$ is a refinement of $\mathcal{G}_-$. We hence denote $\mathcal{G}_-$ being a coarsening of $\mathcal{G}$ by $\mathcal{G}_- \leq \mathcal{G}$ or equivalently $\mathcal{G} \geq \mathcal{G}_-$ as in Definition 3.1.7.

In adaptive algorithms, we are particularly interested in successively modifying a given initial triangulation $\mathcal{G}_{\text{init}}$. For these modifications, we may employ refinement and coarsening. Since the module REFINE produces conforming triangulations, its successive application to $\mathcal{G}_0 := \mathcal{G}_{\text{init}}$ generates a sequence $\{\mathcal{G}_k\}_{k \geq 0}$ of conforming triangulations $\mathcal{G}_k \in \mathbb{G}$ and particularly $\mathcal{G}_k \geq \mathcal{G}_0$ for all $k \geq 0$. Equally, above Definition 3.1.11 implies that a *coarsening* $\mathcal{G}_{k+1}$ of such a $\mathcal{G}_k$ still is a refinement of $\mathcal{G}_{\text{init}}$, i. e., $\mathcal{G}_{k+1} \geq \mathcal{G}_{\text{init}}$. Provided the coarsened mesh $\mathcal{G}_{k+1}$ is conforming, we also have $\mathcal{G}_{k+1} \in \mathbb{G}$.

In analogy to above black box function **REFINE**, we also introduce a function **COARSEN**. Again, we collect the elements to be coarsened in the set $\mathcal{M} \subset \mathcal{G}$ of marked elements and associate an element marker $m(E) \in \mathbb{N}$ to each marked element. In case of coarsening, the element marker $m(E)$ states how many times the element $E \in \mathcal{M}$ may be coarsened. As for refinement, we summarize all relevant information in the set $\mathcal{M}^m \subset \mathcal{G} \times \mathbb{N}$. Given a conforming triangulation $\mathcal{G} \in \mathbb{G}$ and the set $\mathcal{M}^m$ of marked elements and corresponding element markers, we assume we have at hand a function **COARSEN**, which produces a conforming triangulation $\mathcal{G}_- \leq \mathcal{G}$. In this process, all unmarked elements (i. e., elements in $\mathcal{G} \setminus \mathcal{M}$) are left untouched and only elements $E \in \mathcal{M}$ may be coarsened up to $m(E)$ times. More precisely, we have

$$\mathcal{G}_- = \mathsf{COARSEN}(\mathcal{M}^m, \mathcal{G})$$

with $\mathcal{G}_- \leq \mathcal{G}$ and $\mathcal{G} \setminus \mathcal{M} \subset \mathcal{G}_-$.

Note that whereas **REFINE** $(\mathcal{M}^m, \mathcal{G})$ guarantees that all elements in $\mathcal{M}$ are indeed refined, **COARSEN** $(\mathcal{M}^m, \mathcal{G})$ guarantees that all *un*marked elements $\mathcal{G} \setminus \mathcal{M}$ are *not* coarsened. This asymmetry is due to conformity in the following sense: Refining a single element may demand another (possibly unmarked) element being refined to preserve conformity. Likewise, coarsening a single element E_1 may demand another (possibly unmarked) element E_2 being coarsened. As we do not want to allow for unmarked elements to be coarsened, we rather prohibit coarsening of E_1 instead of additionally coarsening E_2. This particularly implies that not all marked elements $E \in \mathcal{M}$ are actually coarsened $m(E)$ times.

In Section 5.1 we review some coarsening strategies and particularly introduce the very flexible multi-coarsening. For implementational details concerning coarsening in general, we refer to [57, Section 3.4].

3.1.2   Finite element spaces

In this section we will make use of the just introduced concept of triangulations to define a suitable finite dimensional subspace $\mathbb{V} \subset H_0^1(\Omega)$, which will be the foundation of the discretization of Problem (2.1) introduced in Section 3.2. Since in Chapter 6 we will need a finite dimensional subspace $\widetilde{\mathbb{V}} \subset H^1(\Omega)$ *without* zero boundary, we introduce the following definitions first with respect to $\widetilde{\mathbb{V}} \subset H^1(\Omega)$ before considering the space with zero boundary $\mathbb{V} \subset H_0^1(\Omega)$.

More general concepts and definitions on finite elements can be found in [11, 15, 9, 40]. For implementational aspects of finite elements we refer to [57].

Definition 3.1.12 (Abstract finite element spaces)
Let $\mathcal{G}$ be a conforming triangulation of Ω and for any $E \in \mathcal{G}$, let $\mathbb{P}(E) \subset C^1(E)$ be a finite dimensional function space on E. An abstract *finite element space* on the triangulation $\mathcal{G}$ is then defined as

$$\widetilde{\mathbb{V}}(\mathcal{G}) := \{V \in C(\overline{\Omega}) \mid V_{|E} \in \mathbb{P}(E) \text{ for all } E \in \mathcal{G}\}.$$

Moreover, we define an abstract *finite element space with zero boundary values* on the triangulation $\mathcal{G}$ as

$$\mathbb{V}(\mathcal{G}) := \{V \in C(\overline{\Omega}) \mid V_{|E} \in \mathbb{P}(E) \text{ for all } E \in \mathcal{G} \text{ and } V_{|\partial\Omega} = 0\}.$$

We point out, that the condition $V_{|\partial\Omega} = 0$ in the definition of a finite element space with zero boundary is tailored to the specific Problem (2.1) considered in this work in the sense that it accounts for the homogeneous Dirichlet boundary condition. Hence, we will use a finite element space of this type throughout this chapter as well as Chapters 4 and 5, which also deal with problem (2.1). Apart from that, the problem considered in Chapter 6 does not involve Dirichlet boundary conditions and hence, we employ a finite element space without zero boundary in that chapter. Moreover, we anticipate the definition of Lagrange finite element spaces and mention that the finite dimensional function space $\mathbb{P}(E)$ typically consists of polynomials up to a certain fixed order.

The following lemma states that for a conforming triangulation $\mathcal{G}$ of Ω, the finite element space $\widetilde{\mathbb{V}}(\mathcal{G})$ is indeed a subspace of $H^1(\Omega)$ and $\mathbb{V}(\mathcal{G})$ additionally is a subspace of $H_0^1(\Omega)$. This property eventually leads to a *conforming* discretization of (2.1), see Section 3.2.

Lemma 3.1.13 (H^1- and H_0^1-conformity)
Let $\mathcal{G}$ be a conforming triangulation of Ω. Then we have $\widetilde{\mathbb{V}}(\mathcal{G}) \subset H^1(\Omega)$ and $\mathbb{V}(\mathcal{G}) \subset H_0^1(\Omega)$.

Proof. Let $V \in \widetilde{\mathbb{V}}(\mathcal{G})$. To show that $V \in H^1(\Omega)$, we have to show that V possesses a weak derivative. For this purpose, let $\varphi \in C_0^\infty(\Omega)$ be an arbitrary test function. Then, for any $i \in \{1, \ldots, d\}$ there holds by integration by parts

$$\int_\Omega V \frac{\partial \varphi}{\partial x_i} \, dx = \sum_{E \in \mathcal{G}} \int_E V \frac{\partial \varphi}{\partial x_i} \, dx$$

$$= \sum_{E \in \mathcal{G}} \int_E -\frac{\partial V}{\partial x_i} \varphi \, dx + \sum_{E \in \mathcal{G}} \sum_{S \subset \partial E} \int_S V \varphi \, \eta_{E,i} \, ds.$$

Here, we used the notation $\eta_{E,i}$ to indicate the i-th component of the unit exterior normal to ∂E. Further, we observe that the boundary term on the right hand side vanishes: If $S \subset \partial\Omega$, we have $\varphi_{|S} = 0$. In case of S being an interior side, there exists a unique element $E' \in \mathcal{G}$ such that $S = E \cap E'$ and $\eta_{E',i} = -\eta_{E,i}$. Since V particularly is continuous, the boundary term also vanishes in this case.

Consequently, $w \in L^\infty(\Omega)$ given by $w_{|E} = \frac{\partial V_{|E}}{\partial x_i}$ for all $E \in \mathcal{G}$ is the i-th weak derivative of V and we conclude $V \in H^1(\Omega)$. If additionally $V \in \mathbb{V} \subset \widetilde{\mathbb{V}}$, we also have $V_{|\partial\Omega} = 0$ by definition and we conclude $V \in H_0^1(\Omega)$ in this case. $\qquad\square$

We now aim at introducing *linear Lagrange finite elements*, which we employ in the computational realizations of Chapters 5 and 6. As indicated above, we use as a finite dimensional function space on a given element $E \in \mathcal{G}$ the space $\mathbb{P}_1(E)$ of polynomials of degree one. Moreover, we wish to have at hand a convenient way for characterizing those local polynomials. For this purpose, linear Lagrange finite elements employ evaluations of the polynomials in the vertices of the grid $\mathcal{G}$ underlying the finite element space.

Remark 3.1.14 (General finite elements)
In the formal definition of a finite element, which can be found in [11], the mentioned characterization of the local functions from $\mathbb{P}(E)$ is realized by defining a particular basis $\{N_1, \ldots, N_k\}$ of the dual space $\mathbb{P}'(E)$. Any function $p \in \mathbb{P}(E)$ can then be identified with the the reals $N_1(p), \ldots, N_k(p)$. Particularly in case of higher order Lagrange finite elements, the basis of $\mathbb{P}'(E)$ is given by functionals which evaluate $p \in \mathbb{P}(E)$ in additional points $\lambda \in E$, the so-called Lagrange nodes.

In the following theorem, we define linear Lagrange elements and state basic properties. A proof of these can be found in most books on finite elements, e. g., [11].

Theorem 3.1.15 (Lagrange elements)
Let δ_{ij} denote the Kronecker symbol. We define

a) the space of linear finite elements

$$\widetilde{\mathbb{V}}^1(\mathcal{G}) := \{V \in C(\overline{\Omega}) \mid V_{|E} \in \mathbb{P}_1(E) \text{ for all } E \in \mathcal{G}\} \subset H^1(\Omega).$$

Any function $V \in \widetilde{\mathbb{V}}^1(\mathcal{G})$ is uniquely defined by its values in the vertices $\mathcal{V}_\mathcal{G} = \{v_1, \ldots, v_N\}$. The nodal basis of $\widetilde{\mathbb{V}}^1(\mathcal{G})$ is given by the set of functions $\{\varphi_1, \ldots, \varphi_N\} \subset \widetilde{\mathbb{V}}^1(\mathcal{G})$ with

$$\varphi_i(v_j) := \delta_{ij} \qquad \text{for all } i, j = 1, \ldots, N.$$

b) the space of linear finite elements with zero boundary values

$$\mathbb{V}^1(\mathcal{G}) := \{V \in C(\overline{\Omega}) \mid V_{|E} \in \mathbb{P}_1(E) \text{ for all } E \in \mathcal{G} \text{ and}$$
$$V_{|\partial\Omega} = 0\} \subset H_0^1(\Omega).$$

Any function $V \in \mathbb{V}^1(\mathcal{G})$ is uniquely defined by its values in the interior vertices $\mathring{\mathcal{V}}_\mathcal{G} = \{v_1, \ldots, v_{\mathring{N}}\}$. The nodal basis of $\mathbb{V}^1(\mathcal{G})$ is given by the set of functions $\{\varphi_1, \ldots, \varphi_{\mathring{N}}\} \subset \mathbb{V}^1(\mathcal{G})$ with

$$\varphi_i(v_j) := \delta_{ij} \qquad \text{for all } i, j = 1, \ldots, \mathring{N}.$$

A very important feature of the defined Lagrange finite element spaces is that they are *nested*. This is, for a conforming triangulation $\mathcal{G}$ and a refinement $\mathcal{G}_+ \geq \mathcal{G}$ we have $\widetilde{\mathbb{V}}(\mathcal{G}) \subset \widetilde{\mathbb{V}}(\mathcal{G}_+)$ as well as $\mathbb{V}(\mathcal{G}) \subset \mathbb{V}(\mathcal{G}_+)$.

In the following theorem, we present the well-known Lagrange interpolation operator. Even though this interpolation is limited to functions allowing for point-wise evaluations, i. e., continuous functions, it is an important tool in computational implementations of finite elements. In particular, we use the Lagrange interpolation operator in Section 5.1 to transfer finite element functions from one finite element space to another.

Theorem 3.1.16 (Lagrange interpolation)

Let $m \leq 2$ such that $m - \frac{d}{p} > 0$. Then the Lagrange interpolation *operator $\mathcal{I}_{\widetilde{\mathbb{V}}^1(\mathcal{G})} \in L(W^{m,p}(\Omega), \widetilde{\mathbb{V}}^1(\Omega))$, which is uniquely defined via*

$$(\mathcal{I}_{\widetilde{\mathbb{V}}^1(\mathcal{G})}g)(v) := g(v) \qquad \text{for all } v \in \mathcal{V}_{\mathcal{G}},$$

satisfies the following interpolation estimates for $0 \leq n \leq m$ for all $g \in W^{m,p}(\Omega)$:

$$\left| \mathcal{I}_{\widetilde{\mathbb{V}}^1(\mathcal{G})}g - g \right|_{W^{n,p}(\Omega)} \lesssim \left(\sum_{E \in \mathcal{G}} h_E^{p(m-n)} |g|_{W^{m,p}(\Omega)}^p \right)^{1/p}$$

Moreover, the Lagrange interpolation *operator $\mathcal{I}_{\mathbb{V}^1(\mathcal{G})} \in L(W_0^{m,p}(\Omega), \mathbb{V}^1(\mathcal{G}))$, which is uniquely defined via*

$$(\mathcal{I}_{\mathbb{V}^1(\mathcal{G})}g)(v) := g(v) \qquad \text{for all } v \in \mathring{\mathcal{V}}_{\mathcal{G}},$$

satisfies the analogue estimate. The constant in $\lesssim$ depends on properties of the associated interpolation operator on the reference simplex $\hat{E}$ and the shape regularity $\sigma_{\mathcal{G}}$ of $\mathcal{G}$.

Proof. See e. g., [11, Chapter 4.4] $\qquad\qquad\qquad\qquad\qquad$ □

Whereas the restriction $m \leq 2$ in Theorem 3.1.16 is due to the fact that we only regard *linear* finite elements, the assumption $m - \frac{d}{p} > 0$ ensures by Sobolev's Embedding Theorem (see e. g., [69, 3, 66]) that the considered functions $g \in W^{m,p}(\Omega)$ can be interpreted as continuous functions. Hence, the evaluation $g(v)$ in the vertices $v \in \mathcal{V}_{\mathcal{G}}$ is well-defined. However, for $d \geq 2$, this continuity asks for regularity exceeding $H_0^1(\Omega)$. To avoid such a restriction, we resort to a quasi-interpolation operator like the Clément or Scott-Zhang interpolation, cf. [17] and [58]. In the following theorem, we cite an estimate for the corresponding interpolation error, a proof can be found e. g., in [11, 58, 40].

Theorem 3.1.17 (Quasi-interpolation)

Let $m \leq 2$ and $1 \leq p \leq \infty$. Moreover, let $0 \leq n \leq m$ and $1 \leq q \leq \infty$ such that $m - d/p > n - d/q$. Then there exists an operator $\mathcal{I}_{\widetilde{\mathbb{V}}^1(\mathcal{G})} : L^1(\Omega) \to \widetilde{\mathbb{V}}^1(\mathcal{G})$ such that for every $E \in \mathcal{G}$

$$\left| \mathcal{I}_{\widetilde{\mathbb{V}}^1(\mathcal{G})}v - v \right|_{W^{n,q}(E)} \lesssim h_E^{(m-d/p)-(n-d/q)} |v|_{W^{m,p}(\tilde{\omega}_E)}.$$

Moreover, for $m \geq 1$ there exists an operator $\mathcal{I}_{\mathbb{V}^1(\mathcal{G})} : W_0^{1,1}(\Omega) \to \mathbb{V}^1(\mathcal{G})$ which satisfies the same estimate. Both operators are invariant on $\widetilde{\mathbb{V}}^1(\mathcal{G})$ respectively $\mathbb{V}^1(\mathcal{G})$. The constant hidden in $\lesssim$ depends on the shape regularity $\sigma_{\mathcal{G}}$ and the dimension d.

Apart from Chapter 6, where we explicitly point out the notation, we conveniently use the notation

$$\mathbb{V}(\mathcal{G}) := \mathbb{V}^1(\mathcal{G})$$

for a finite element space with zero boundary values. In particular, we will *not* employ a finite element space *without* zero boundary values throughout this chapter as well as Chapters 4 and 5. With this in mind, no notational confusion should arise.

3.2 Space time discretization

We now aim at deriving an approximation to the solution of (2.1). To that end, we consider the weak formulation (2.16). We notice, that the weak solution $u \in \mathbb{W}(0, T)$ is depending both on space *and* time which requires an approximation in both variables. Solely regarding the time dependency, equation (2.16) has the structure of an *ordinary* differential equation

$$\frac{d}{dt}u(t) = F(t, u(t)) \qquad \text{for } t \in (0, T)$$
$$u(0) = u_0.$$

On the other hand, completely neglecting the time dependency of u — which particularly implies $u' = 0$ — equation (2.16) constitutes the weak formulation of a *partial* differential equation

$$\mathcal{B}(u, v) = \langle f, v \rangle \qquad \text{for all } v \in H_0^1(\Omega),$$

with a bilinear form $\mathcal{B}(\cdot, \cdot)$ associated to an elliptic second order differential operator. This fundamental disparity between time and space dependency will be reflected in different types of discretization: First, we discretize (2.16) in space using finite elements. This eventually leads to an ordinary differential equation in an $\mathbb{R}^N$, where N is the dimension of the employed finite element space.

To solve this ODE, we then employ a time stepping method, in particular the implicit Euler method. This procedure of first discretizing in space yielding an ordinary differential equation in $\mathbb{R}^N$ is called *method of lines*. However, for the full space time discretization that we are ultimately aiming for, this is equivalent to the so-called Rothe's method, which first discretizes in time before employing a spatial discretization.

3.2.1 Spatial discretization

In order to obtain a spatial discretization of (2.16), we proceed analogously to the derivation of the Galerkin approximation of Section 2.2. In fact, the only difference to the statement of (Gal-E) is the choice of the finite dimensional space. Whereas in Section 2.2 we aimed for a representation of the solution $u_N(t) \in \mathbb{V}_N$, which was suitable for a straightforward analysis of the existence and uniqueness of a solution of (2.16), the focus now is on applicability. We therefore assume we have at hand a conforming triangulation $\mathcal{G}$ of Ω and we choose the finite element space $\mathbb{V}(\mathcal{G})$ as finite dimensional subspace of $H_0^1(\Omega)$, see Lemma 3.1.13. Hence, in the *finite element spatial discretization* of (2.16) we are looking for a function $U : [0, T] \to \mathbb{V}(\mathcal{G})$ of the form

$$U(t) := \sum_{k=1}^{N} \alpha_k(t)\varphi_k,$$

where φ_k are the basis functions of $\mathbb{V}(\mathcal{G})$ and α_k, $k = 1, \ldots, N$ are time dependent coefficients. Like in Section 2.2, we again demand that U fulfills

$$\langle U'(t), V \rangle + \mathcal{B}(U(t), V) = \langle f(t), V \rangle \quad \text{for all } V \in \mathbb{V}(\mathcal{G}),$$
$$\text{f.a.e. } t \in (0, T) \quad \text{(3.1a)}$$
$$U(0) = U_0. \quad \text{(3.1b)}$$

Here, $U_0 \in \mathbb{V}(\mathcal{G})$ is an arbitrary approximation of $u_0 \in L^2(\Omega)$ in the finite element space $\mathbb{V}(\mathcal{G})$. Existence and uniqueness of a function U of the above form satisfying (3.1) follows as in Theorem 2.2.2.

Remark 3.2.1 (Approximation of initial data)
The natural choice for the approximation $U_0 \in \mathbb{V}(\mathcal{G})$ of the initial

data $u_0 \in L^2(\Omega)$ is the L^2-projection $\Pi_{\mathbb{V}(\mathcal{G})} u_0$. To state this more precisely, the linear mapping $\Pi_{\mathbb{V}(\mathcal{G})} : L^2(\Omega) \to \mathbb{V}(\mathcal{G})$ is characterized by

$$\langle \Pi_{\mathbb{V}(\mathcal{G})} v \, , \, V \rangle = \langle v \, , \, V \rangle \qquad \text{for all } V \in \mathbb{V}(\mathcal{G}).$$

On the other hand, the subsequent analysis does not require this exact choice of the initial approximation and hence, different choices are valid as well. From an implementational point of view, it is more convenient to use an interpolation operator. In many practical applications, the initial data possesses more than L^2-regularity and often provides at least piecewise continuity allowing the use of pointwise interpolants. The computational effort for computing such interpolations is much less than for computing a projection.

3.2.2 Full space time discretization

For the discretization in time, we employ the well-known *implicit Euler* method, which is commonly used in the discretization of ordinary differential equations. We divide the time interval $(0, T)$ into subintervals I_n and approximate the temporal derivative on each of those subintervals. To be more precise, let

$$0 = t_0 < t_1 < \cdots < t_N = T$$

be a partition of the time interval $(0, T)$ and set $I_n := [t_{n-1}, t_n]$. Conveniently, we also define the local *time step size*

$$\tau_n := |I_n| = t_n - t_{n-1} \qquad \text{for } n = 1, \ldots, N.$$

With this notation, the *implicit Euler Galerkin* discretization of equation (2.16) reads: For given $U_0 \in \mathbb{V}(\mathcal{G})$ find $U_n \in \mathbb{V}(\mathcal{G})$, $n = 1, \ldots, N$ such that

$$\frac{1}{\tau_n} \langle U_n - U_{n-1} \, , \, V \rangle + \mathcal{B}(U_n \, , \, V) = \langle f_n \, , \, V \rangle \qquad \text{for all } V \in \mathbb{V}(\mathcal{G}).$$
$$(3.2)$$

Besides replacing the temporal derivative U' in (3.1a) by the difference quotient $\frac{1}{\tau_n}(U_n - U_{n-1})$, the *implicit* Euler method evaluates the involved time dependent functions *implicitly*, i.e., at time t_n. Hence, the term $\mathcal{B}(U(t) \, , \, V)$ of (3.1a) is straightforwardly replaced by $\mathcal{B}(U_n \, , \, V)$. Since the right hand side function

$f \in L^2(0, T; L^2(\Omega))$ is not necessarily continuous with respect to time, an evaluation $f(t_n)$ lacks justification. We therefore retreat to defining $f_n \in L^2(\Omega)$ as the mean value of f on the n-th subinterval I_n, i. e.,

$$f_n := \frac{1}{\tau_n} \int_{I_n} f(s) \, ds \in L^2(\Omega).$$

With this definition, equation (3.2) constitutes a full discretization of the weak formulation (2.16). It also contains some notion of adaptivity since the time step sizes τ_n may be chosen individually for each time step. An adaptive algorithm can make use of this freedom of choice to adjust the local time step size τ_n in the n-th time step to the current temporal behavior of the solution. On the other hand, equation (3.2) employs the same finite element space $\mathbb{V}(\mathcal{G})$ in every time step and hence does not provide any adaptivity with respect to space. To resolve this issue, we allow for an individual triangulation $\mathcal{G}_n$ and associated finite element space $\mathbb{V}_n := \mathbb{V}(\mathcal{G}_n)$ in each time step n. With this concept, the *fully adaptive* implicit Euler Galerkin discretization reads: For given $U_0 \in \mathbb{V}_0$ find $U_n \in \mathbb{V}_n$, $n = 1, \ldots, N$, such that

$$\frac{1}{\tau_n}\langle U_n - U_{n-1}, V \rangle + \mathcal{B}(U_n, V) = \langle f_n, V \rangle \qquad \text{for all } V \in \mathbb{V}_n.$$
$$\text{(EG)}$$

Remark 3.2.2
In case of nested finite element spaces $\mathbb{V}_{n-1} \subset \mathbb{V}_n$, we have $U_{n-1} \in \mathbb{V}_n$ and (EG) may resolve the full information of U_{n-1}. Opposed to that, in case of non-nested finite element spaces with $U_{n-1} \notin \mathbb{V}_n$, (EG) implicitly employs the L^2-projection to transfer U_{n-1} to $\mathbb{V}_n$ in the sense that the first term of (EG) satisfies

$$\frac{1}{\tau_n}\langle U_n - U_{n-1}, V \rangle = \frac{1}{\tau_n}\langle U_n - \Pi_n U_{n-1}, V \rangle \qquad \text{for all } V \in \mathbb{V}_n.$$

The most interesting question arising from (EG) is how to choose the individual time step sizes τ_n as well as the finite element spaces $\mathbb{V}_n$. Before addressing this in Chapter 4, we note that employing a scheme like (EG), we transform a *time dependent* problem (of the form (2.16)) into a sequence of *stationary* problems. In particular, rewriting (EG) as

$$\mathcal{B}(U_n, V) + \frac{1}{\tau_n}\langle U_n, V\rangle = \langle f_n + \frac{1}{\tau_n}U_{n-1}, V\rangle \qquad \text{for all } V \in \mathbb{V}_n$$

and keeping in mind that in the n-th time step the previous solution $U_{n-1} \in \mathbb{V}_{n-1}$ is already known, emphasizes that (EG) computes an (adaptive) approximation to the solution $u_n \in H_0^1(\Omega)$ of the *elliptic problem*

$$\mathcal{B}(u_n, v) + \frac{1}{\tau_n}\langle u_n, v\rangle = \langle f_n + \frac{1}{\tau_n}U_{n-1}, v\rangle \qquad \text{for all } v \in H_0^1(\Omega).$$

We briefly comment on how the discrete solution $U_n \in \mathbb{V}_n$ from a finite dimensional space $\mathbb{V}_n$ can be computed in the n-th time step.

Remark 3.2.3 (Computational solution)
Since we are dealing with a linear problem in (EG), we consider a basis $\{\varphi_1, \ldots, \varphi_M\}$ of the finite dimensional space $\mathbb{V}_n$, $\dim \mathbb{V}_n =: M = M(n)$. Then equation (EG) can be equivalently expressed as

$$\mathcal{B}(U_n, \varphi_i) + \frac{1}{\tau_n}\langle U_n, \varphi_i\rangle = \langle f_n + \frac{1}{\tau_n}U_{n-1}, \varphi_i\rangle \qquad \text{for } i = 1, \ldots, M.$$

Representing the solution $U_n \in \mathbb{V}_n$ as a linear combination of the basis vectors, i.e., $U_n = \sum_{j=1}^{M} \alpha_j\varphi_j$ with coefficients $\alpha_j \in \mathbb{R}$, $j = 1, \ldots, M$, and exploiting the linearity of $\mathcal{B}$, we see that the Galerkin solution is characterized as

$$\sum_{j=1}^{M} \alpha_j \left(\mathcal{B}(\varphi_j, \varphi_i) + \frac{1}{\tau_n}\langle \varphi_j, \varphi_i\rangle\right) = \langle f_n + \frac{1}{\tau_n}U_{n-1}, \varphi_i\rangle$$

$$\text{for } i = 1, \ldots, M. \quad (3.3)$$

Conveniently, we may define the matrix

$$\mathbf{B} := (\mathcal{B}(\varphi_j, \varphi_i) + \frac{1}{\tau_n}\langle \varphi_j, \varphi_i\rangle)_{i,j=1,\ldots,M} \in \mathbb{R}^{M\times M}$$

as well as the vectors $\mathbf{f} := (\langle f_n + \frac{1}{\tau_n}U_{n-1}, \varphi_i\rangle)_{i=1,\ldots,M}^{\mathsf{T}} \in \mathbb{R}^M$ and $\alpha := (\alpha_i)_{i=1,\ldots,M}^{\mathsf{T}} \in \mathbb{R}^M$ to rewrite equation (3.3) as the linear equation

$$\mathbf{B}\,\alpha = \mathbf{f}.$$

This M-dimensional linear equation can be solved by employing computational linear algebra, e.g., modern iterative solvers.

This motivates a brief digression into elliptic problems.

3.3 Adaptive methods for elliptic problems

In this section, we consider *stationary* partial differential equations of the form

$$\mathcal{L}u = f \qquad \text{in } \Omega,$$
$$u = 0 \qquad \text{on } \partial\Omega, \tag{3.4}$$

where $\mathcal{L}$ is a general elliptic operator of the same form as in (2.2) in the beginning of Chapter 2 and $f \in L^2(\Omega)$. We point out that when considering an elliptic problem emerging from above discretization of a parabolic equation, the current time step size enters the zero order term. After establishing the existence of a unique solution u of (3.4) and employing the Galerkin discretization to give rise to an approximation of u, we briefly state a well-known a priori estimate. Most beneficial with respect to the considered time dependent problems, however, we review a posteriori error analysis for elliptic problems and particularly focus on the upper bound as we may reuse it in the context of parabolic problems in Section 3.4. Furthermore, we review adaptivity for stationary problems by stating an adaptive algorithm which is guaranteed to terminate within any given tolerance by a basic convergence result by Morin, Siebert, and Veeser, cf. [36]. This will be an essential component in the convergence analysis for parabolic problems in Section 4.5.

We start with recalling the bilinear form $\mathcal{B}$, which we introduced in Definition 2.1.1 as

$$\mathcal{B} : H_0^1(\Omega) \times H_0^1(\Omega) \to \mathbb{R}$$

$$\mathcal{B}(v\,,\,w) := \int_\Omega \mathbf{A}\nabla v \cdot \nabla w + \mathbf{b} \cdot \nabla v\, w + c\, v\, w\, dx,$$

and define the notion of a weak solution of (3.4).

Definition 3.3.1 (Weak solution)
A function $u \in H_0^1(\Omega)$ is called *weak solution* of (3.4) provided

$$\mathcal{B}(u\,,\,v) = \langle f\,,\,v \rangle \qquad \text{for all } v \in H_0^1(\Omega). \tag{3.5}$$

Remark 3.3.2 (Strong and weak solutions)
Analogously to the proceeding in Remark 2.1.12, we consider the relation of weak and strong solutions of the elliptic problem (3.4). Assume that (3.4) possesses a classical solution, i.e., a function u regular enough such that the differential operator can be applied to u using classic derivatives. In particular, for the employed operator $\mathcal{L}$ this means there exists a function $u \in C^2(\Omega) \cap C(\overline{\Omega})$ such that equation (3.4) holds. We may then multiply (3.4) with an arbitrary test function $v \in H_0^1(\Omega)$, integrate over Ω and apply integration by parts to find

$$\int_\Omega \mathbf{A}\nabla u \cdot \nabla v + \mathbf{b} \cdot \nabla u \, v + c \, u \, v \, dx = \int_\Omega f \, v \, dx.$$

Hence, we conclude that any strong solution u is also a weak solution in the sense of Definition 3.3.1. Conversely, if a weak solution u possesses sufficient regularity, we may revert the integration by parts to derive from (3.5)

$$\int_\Omega \mathcal{L}u \, v \, dx = \int_\Omega f \, v \, dx \qquad \text{for all } v \in H_0^1(\Omega).$$

Further, we conclude by the Fundamental Lemma of calculus of variations that u is indeed a classical solution of (3.4) since the regularity $u \in C(\overline{\Omega})$ and $u \in H_0^1(\Omega)$ also implies that the boundary condition is satisfied.

3.3.1 Existence and uniqueness

In order to establish the existence of a unique weak solution of (3.4), we employ the Lax-Milgram Theorem stated below. This theorem is an extension of the Riesz Representation Theorem to non-symmetric bilinear forms and can be found in most books on functional analysis, we exemplarily refer to [69, 3, 66].

Theorem 3.3.3 (Lax-Milgram Theorem)
Let $\mathbb{X}$ be a real Hilbert space with dual space $\mathbb{X}'$. Assume that $B : \mathbb{X} \times \mathbb{X} \to \mathbb{R}$ is a bilinear mapping, for which there exist constants $C_1, C_2 > 0$ such that for all $v, w \in \mathbb{X}$

$$|B(v, w)| \leq C_1 \, \|v\|_{\mathbb{X}} \, \|w\|_{\mathbb{X}} \qquad (continuity),$$
$$C_2 \, \|v\|^2 \leq B(u, u) \qquad (coercivity).$$

Then, for any $f \in \mathbb{X}'$ there exists a uniquely determined $u \in \mathbb{X}$ such that

$$B(u, v) = \langle f, v \rangle_{\mathbb{X}' \times \mathbb{X}} \quad \text{for all } v \in \mathbb{X}.$$

In view of the weak formulation given in (3.5), we wish to apply the Lax-Milgram Theorem in the Hilbert space $H_0^1(\Omega)$ and use $\mathcal{B}(\cdot, \cdot)$ as the bilinear mapping. First, we notice that $f \in L^2(\Omega)$ can indeed be interpreted as functional $f \in H^{-1}(\Omega)$ by defining

$$\langle f, v \rangle_{H^{-1}(\Omega) \times H_0^1(\Omega)} := \langle f, v \rangle_{L^2(\Omega)} \quad \text{for all } v \in H_0^1(\Omega) \subset L^2(\Omega).$$

From the definition of the operator norm $\| \cdot \|_*$ as well as Cauchy-Schwarz and Poincaré inequality we thus have

$$\|f\|_* = \sup_{v \in H_0^1(\Omega)} \frac{\langle f, v \rangle_{L^2(\Omega)}}{\|v\|_\Omega} \leq \|f\|_{L^2(\Omega)} \sup_{v \in H_0^1(\Omega)} \frac{\|v\|_{L^2(\Omega)}}{\|v\|_\Omega}$$
$$\leq \mathcal{C}_P \|f\|_{L^2(\Omega)}.$$

It remains to verify the hypothesis of the Lax-Milgram Theorem regarding continuity and coercivity of $\mathcal{B}(\cdot, \cdot)$. For this purpose, we recall Lemma 2.1.3, which particularly states

$$|\mathcal{B}(v, w)| \leq \mathcal{C}_B \|v\|_\Omega \|w\|_\Omega \quad \text{for all } v, w \in H_0^1(\Omega)$$

and as the energy norm is equivalent to the standard $H_0^1(\Omega)$-norm, this gives the desired continuity. Regarding the coercivity, we observe from the equivalence of the norms $\| \cdot \|_{H_0^1(\Omega)}$ and $\| \cdot \|_\Omega$ as well as the definition of the energy norm

$$\mathcal{C}_-^2 \|v\|_{H_0^1(\Omega)}^2 \leq \|v\|_\Omega^2 = \mathcal{B}(v, v).$$

Hence, we verified the hypothesis and applying the Lax-Milgram Theorem directly implies existence of a unique weak solution $u \in H_0^1(\Omega)$ of (3.4).

Corollary 3.3.4 (Existence and uniqueness)
There exists a unique weak solution $u \in H_0^1(\Omega)$ of (3.4) satisfying

$$\|u\|_\Omega \leq \|f\|_*.$$

Proof. We only need to establish the energy estimate. Therefore, we choose $v := u \in H_0^1(\Omega)$ as a test function in equation (3.5) to find

$$\|u\|_\Omega^2 = \mathcal{B}(u, u) = \langle f, u \rangle \leq \|f\|_* \|u\|_\Omega,$$

which proves the claim. $\qquad\square$

Remark 3.3.5 (Existence and uniqueness for symmetric operators)

For *symmetric* bilinear forms, which commonly appear if no first order term is involved in the differential equation (3.4), existence and uniqueness of a solution can be established more directly by noting that in this case, $\mathcal{B}(\cdot, \cdot)$ defines an alternate scalar product on $H_0^1(\Omega)$. Hence, the Riesz Representation Theorem (cf. [69, 3, 66]) can be used on the Hilbert space $(H_0^1(\Omega), \mathcal{B}(\cdot, \cdot))$ and directly yields the existence of a unique weak solution.

3.3.2 Galerkin discretization

In order to obtain a computable approximation of the solution of (3.4), we proceed analogously as in Section 3.2.1 and restrict the infinite dimensional ansatz and test space $H_0^1(\Omega)$ to a finite dimensional space. In particular, we choose a finite element space $\mathbb{V} = \mathbb{V}(\mathcal{G})$ for some conforming triangulation $\mathcal{G}$.

Definition 3.3.6 (Discrete solution)
Let $\mathbb{V} \subset H_0^1(\Omega)$ be a finite dimensional subspace. Then a solution $U \in \mathbb{V}$ to

$$\mathcal{B}(U, V) = \langle f, V \rangle \qquad \text{for all } V \in \mathbb{V} \tag{3.6}$$

is called Galerkin solution.

It follows directly that there exists a unique Galerkin solution of (3.6) since we are using a conforming finite element space $\mathbb{V} \subset H_0^1(\Omega)$. This conformity guarantees that the bilinear form $\mathcal{B}$ is also continuous and coercive on the subspace $\mathbb{V}$ and hence, by the same arguments as in the continuous case in Section 3.3.1, the Lax-Milgram Theorem establishes the existence of a unique solution $U \in \mathbb{V}$. In the following corollary, we state a stability estimate for the Galerkin solution which follows analogously to Corollary 3.3.4 by using $V := U \in \mathbb{V}$ as a test function in (3.6)

Corollary 3.3.7 (Discrete solution)
There exists a unique Galerkin solution $U \in \mathbb{V}$ of (3.6). This solution satisfies the stability estimate

$$\|U\|_\Omega \leq \|f\|_* .$$

3.3.3   A priori error analysis

We will now justify the above definition of a Galerkin solution by presenting bounds for the discretization error. Since we consider a conforming discretization with $\mathbb{V} \subset H_0^1(\Omega)$, the approximation $U \in \mathbb{V}$ is also a function in $H_0^1(\Omega)$. Thus the discretization error $u - U$ may be measured in any norm on $H_0^1(\Omega)$. Particularly, we are interested in the energy norm $\|u - U\|_\Omega$.

The essential tool for the proof of Céa's Lemma below is the following *Galerkin orthogonality* which is an essential property of Galerkin solutions. The solution $u \in H_0^1(\Omega)$ satisfies

$$\mathcal{B}(u, v) = \langle f, v \rangle \qquad \text{for all } v \in H_0^1(\Omega) \tag{3.7}$$

whereas the Galerkin solution $U \in \mathbb{V}$ satisfies the same equation on a subspace,

$$\mathcal{B}(U, V) = \langle f, V \rangle \qquad \text{for all } V \in \mathbb{V} \subset H_0^1(\Omega). \tag{3.8}$$

Choosing $v := V \in \mathbb{V} \subset H_0^1(\Omega)$ in (3.7) and subtracting (3.8), we derive the Galerkin orthogonality

$$\mathcal{B}(u - U, V) = 0 \qquad \text{for all } V \in \mathbb{V}. \tag{3.9}$$

Lemma 3.3.8 (Céa's Lemma)
Let $u \in H_0^1(\Omega)$ be the weak solution of (3.5). Then the Galerkin solution $U \in \mathbb{V}$ of (3.6) is the quasi best approximation of u in $\mathbb{V}$ with respect to the energy norm, i. e.,

$$\|u - U\|_\Omega \leq C_\mathcal{B} \min_{V \in \mathbb{V}} \|u - V\|_\Omega .$$

Proof. Using this Galerkin orthogonality (3.9) as well as Cauchy-Schwarz inequality and continuity of $\mathcal{B}$, we find for arbitrary $V \in \mathbb{V}$

$$\|u - U\|_\Omega^2 = \mathcal{B}(u - U, u - U) = \mathcal{B}(u - U, u - V)$$
$$\leq C_\mathcal{B} \|u - U\|_\Omega \|u - V\|_\Omega .$$

Hence, since $V \in \mathbb{V}$ was arbitrary and $\mathbb{V}$ is closed, this implies the assertion. □

Céa's Lemma suggests that the Galerkin solution $U \in \mathbb{V}$ is the quasi best approximation with respect to the energy norm. This implies on the other hand that an explicit estimate of the error $\|u - U\|_{\Omega}$ can be given by (cleverly) choosing a function $\mathcal{I}u \in \mathbb{V}$ with preferably small distance $\|u - \mathcal{I}u\|_{\Omega}$. For that purpose, we employ the Scott-Zhang quasi-interpolation operator and exploiting the interpolation properties of Theorem 3.1.17, we obtain the well-known a priori estimate, cf. [11, 15, 22, 25].

Theorem 3.3.9 (A priori estimate)
Let $u \in H_0^1(\Omega)$ be the weak solution of (3.4) and assume $U \in \mathbb{V}^1(\mathcal{G})$ to be the Galerkin solution in the linear finite element space $\mathbb{V}^1(\mathcal{G})$. Assume further that $u \in H^m(\Omega)$ where $1 < m \leq 2$. Then there holds

$$\|u - U\|_{\Omega} \lesssim \left(\sum_{E \in \mathcal{G}} h_E^{2(m-1)} |u|_{H^m(E)}^2 \right)^{1/2}.$$

The constant hidden in $\lesssim$ depends on properties of $\mathbf{A}$, the shape regularity of $\mathcal{G}$ and the dimension d.

3.3.4 A posteriori error analysis

The a priori analysis presented in Section 3.3.3 provides an error bound which is only of theoretical interest since it involves the *exact* solution $u \in H_0^1(\Omega)$ of (3.5). Since this exact solution is generally unknown, a priori bounds are only able to provide qualitative asymptotical information in form of an error decay during global mesh refinement.

Opposed to that, it would come in useful to have at hand a *computable* error bound. This could be used to ensure that a certain precision in approximating the exact solution is achieved. In order to keep low the computational effort, one might compute an approximation on a relatively coarse grid in a first step. In a second step, the quality of this approximation is evaluated using the computable error bound. If the desired precision is not yet achieved, above a priori analysis suggests refining the mesh globally and recomputing the Galerkin solution. As a direct consequence of

Theorem 3.3.9, iterating this procedure will eventually reduce the discretization error below any given tolerance provided the grid is refined uniformly.

Additionally to providing a computable error bound, the a posteriori error estimate derived in this section also gives information about *local* error contributions. We may use this information to determine how to refine the mesh *cleverly* by identifying regions with huge *local* discretization errors. With such information, global mesh refinement can be replaced by problem adjusted selective refinement.

Throughout this section, we let $u \in H_0^1(\Omega)$ be the weak solution of (3.4) and we assume $U \in \mathbb{V} = \mathbb{V}(\mathcal{G})$ to be the Galerkin solution of (3.6) in the finite element space $\mathbb{V}(\mathcal{G})$.

Upper bound

The upper bound for the discretization error $\|u - U\|_\Omega$ derived in the following is standard in finite element theory and similar estimates can be found in various books (e. g., [64, 2]). However, for completeness and since we directly resort to this estimate also in case of parabolic problems (cf. Section 3.4), we present the derivation.

Definition 3.3.10 (Residual)
We define the *residual* $\mathcal{R} = \mathcal{R}(U, f) \in H^{-1}(\Omega)$ by

$$\langle \mathcal{R}, v \rangle_{H^{-1}(\Omega) \times H_0^1(\Omega)} := \langle f, v \rangle - \mathcal{B}(U, v) \qquad \text{for all } v \in H_0^1(\Omega).$$

In the following, we drop the subscript in dual pairings involving the residual. In light of the weak formulation (3.5), the residual is closely related to the error $u - U$ via

$$\langle \mathcal{R}, v \rangle = \mathcal{B}(u - U, v) \qquad \text{for all } v \in H_0^1(\Omega).$$

The following lemma relates the discretization error in the energy norm to the residual.

Lemma 3.3.11 (Equivalence of error and residual)
The energy norm of the discretization error and the residual are equivalent. More precisely, it holds

$$\|u - U\|_\Omega \leq \|\mathcal{R}\|_* \leq \mathcal{C}_\mathcal{B} \|u - U\|_\Omega$$

with $\mathcal{C}_\mathcal{B}$ denoting the continuity constant from Lemma 2.1.3.

Proof. From the definition of the energy norm and the residual we have

$$\|u - U\|_\Omega^2 = \mathcal{B}(u - U, u - U) = \langle \mathcal{R}, u - U \rangle \le \|\mathcal{R}\|_* \|u - U\|_\Omega$$

and hence $\|u - U\|_\Omega \le \|\mathcal{R}\|_*$. The opposite inequality follows using the definition of the dual norm $\|\cdot\|_*$ and the continuity of $\mathcal{B}$:

$$\|\mathcal{R}\|_* = \sup_{v \in H_0^1(\Omega)} \frac{\langle \mathcal{R}, v \rangle}{\|v\|_\Omega} = \sup_{v \in H_0^1(\Omega)} \frac{\mathcal{B}(u - U, v)}{\|v\|_\Omega} \le C_\mathcal{B} \|u - U\|_\Omega.$$

$\square$

The most important aspect of the equivalence of error and residual is that while the error does involve the exact solution u which is unknown in general, the residual exclusively contains given data and the discrete solution. Lemma 3.3.11 hence enables us (up to a constant) to compute a norm of the error by evaluating the norm of a functional which is completely determined by known quantities. However, computing a dual norm involves finding a supremum in an infinite dimensional space, leaving us with the desire to at least approximate $\|\mathcal{R}\|_*$.

In the following, we estimate this dual norm using standard residual techniques, cf. [64, 2, 40]. Therefore, we decompose the residual into its contributions on each element $E \in \mathcal{G}$:

$$\langle \mathcal{R}, v \rangle = \langle f, v \rangle - \mathcal{B}(U, v)$$

$$= \int_\Omega f\, v - \mathbf{A}\nabla U \cdot \nabla v - \mathbf{b} \cdot \nabla U\, v - c\, U\, v\, dx$$

$$= \sum_{E \in \mathcal{G}} \left(\int_E (f - \mathbf{b} \cdot \nabla U - c\, U)\, v\, dx - \int_E \mathbf{A}\nabla U \cdot \nabla v\, dx \right)$$

Since by definition of the finite element space $\mathbb{V}$, the restriction of $U \in \mathbb{V}$ to any element $E \in \mathcal{G}$ is a polynomial and $\mathbf{A}$ is piecewise Lipschitz over $\mathcal{G}$, we may apply integration by parts to the last term. Realizing that for *linear* finite elements we have $\mathrm{div}(\mathbf{A}\nabla U) = 0$, this yields

$$\langle \mathcal{R}, v \rangle = \sum_{E \in \mathcal{G}} \int_E (f - \mathbf{b} \cdot \nabla U - c\, U)\, v\, dx - \sum_{E \in \mathcal{G}} \int_{\partial E} \mathbf{A}\nabla U \cdot \eta_E\, v\, ds,$$

where η_E denotes the unit exterior normal to the boundary of $E \in \mathcal{G}$. Next, we recall that the trace of $v \in H_0^1(\Omega)$ vanishes on the boundary $\partial\Omega$ and notice that the traces of v match along the edges shared by two neighboring elements. This further yields

$$= \sum_{E \in \mathcal{G}} \int_E (f - \mathbf{b} \cdot \nabla U - c\,U)\, v\, dx + \sum_{S \in \mathring{\mathcal{S}}_\mathcal{G}} \int_S [\![\mathbf{A}\nabla U]\!]\, v\, ds,$$

$$(3.10)$$

where we introduced the notation $[\![\mathbf{A}\nabla U]\!]$ for the jump of $\mathbf{A}\nabla U$ across inter-element sides, i. e.,

$$[\![\mathbf{A}\nabla U]\!]\big|_S := \mathbf{A}\nabla U_{|E_+} \cdot \eta_+ + \mathbf{A}\nabla U_{|E_-} \cdot \eta_-$$

for any interior side $S \in \mathring{\mathcal{S}}_\mathcal{G}$ shared by the elements E_+ and E_-. Here, η_+ and η_- denote the unit interior normals to E_+ and E_- respectively. Note that the jump $[\![\mathbf{A}\nabla U]\!]$ is independent of the particular choice of E_+ and E_-.

The two constituents of (3.10) can be interpreted in the following way:

(1) The first term $\int_E (f - \mathbf{b} \cdot \nabla U - c\,U)\, v\, dx$, which is called *element residual*, is only defined on the individual elements of $\mathcal{G}$. In case of higher order finite elements, the element residual additionally contains the term $\mathrm{div}(\mathbf{A}\nabla U)$ and thus, we see that the element residual accounts for how well the differential equation is fulfilled on the individual elements. Conveniently, we abbreviate

$$R := f - \mathbf{b} \cdot \nabla U - c\,U. \qquad (3.11)$$

(2) The second term on the other hand is only defined on the interior *sides* $\mathring{\mathcal{S}}_\mathcal{G}$ of the triangulation $\mathcal{G}$. From the derivation of equation (3.10) we see that this term arises from the fact that U is not *globally* smooth in Ω but only on the individual elements $E \in \mathcal{G}$. Since $\int_S [\![\mathbf{A}\nabla U]\!]\, v\, ds$ measures the jump of $\mathbf{A}\nabla U$ across the inter-element border $S \in \mathring{\mathcal{S}}_\mathcal{G}$, this term is called *jump residual* and we define $j : \Gamma_\mathcal{G} \to \mathbb{R}$ with

$$j(x) := \begin{cases} [\![\mathbf{A}\nabla U]\!] & \text{for } x \in \Gamma_\mathcal{G} \cap \Omega \\ 0 & \text{for } x \in \partial\Omega. \end{cases} \qquad (3.12)$$

The jump residual accounts for the lack of global smoothness of U.

With this notation, equation (3.10) easily reads

$$\langle \mathcal{R}, v \rangle = \sum_{E \in \mathcal{G}} \int_E R\, v\, dx + \sum_{S \in \mathcal{S}_\mathcal{G}} \int_S j\, v\, ds \qquad \text{for all } v \in H_0^1(\Omega).$$
(3.13)

The key in the derivation of an a posteriori error estimate lies in exploiting the Galerkin orthogonality (3.9) to insert an appropriate interpolation $\mathcal{I}_\mathbb{V} v \in \mathbb{V}$ and exploiting interpolation properties. The Galerkin orthogonality yields $\langle \mathcal{R}, \mathcal{I}_\mathbb{V} v \rangle = \mathcal{B}(u - U, \mathcal{I}_\mathbb{V} v) = 0$ and hence, in combination with equation (3.13) we find

$$\langle \mathcal{R}, v \rangle = \langle \mathcal{R}, v - \mathcal{I}_\mathbb{V} v \rangle = \sum_{E \in \mathcal{G}} \int_E R\,(v - \mathcal{I}_\mathbb{V} v)\, dx$$
$$+ \sum_{S \in \mathcal{S}_\mathcal{G}} \int_S j\,(v - \mathcal{I}_\mathbb{V} v)\, ds.$$

Conveniently, we exploit that any interior side is contained in two elements as well as the fact that j vanishes on non-interior sides to obtain

$$= \sum_{E \in \mathcal{G}} \left(\int_E R\,(v - \mathcal{I}_\mathbb{V} v)\, dx + \frac{1}{2} \int_{\partial E} j\,(v - \mathcal{I}_\mathbb{V} v)\, ds \right)$$

and applying Cauchy-Schwarz inequality further implies

$$\leq \sum_{E \in \mathcal{G}} \Big(\|R\|_{L^2(E)} \, \|v - \mathcal{I}_\mathbb{V} v\|_{L^2(E)}$$
$$+ \frac{1}{2} \|j\|_{L^2(\partial E)} \, \|v - \mathcal{I}_\mathbb{V} v\|_{L^2(\partial E)} \Big).$$
(3.14)

As indicated, we aim at exploiting interpolation properties. To that end, we particularly employ the Scott-Zhang interpolation operator (see [58]) as $\mathcal{I}_\mathbb{V} : H_0^1(\Omega) \to \mathbb{V}$ which satisfies

$$\|v - \mathcal{I}_\mathbb{V} v\|_{L^2(E)} \lesssim h_E \, \|v\|_{H_0^1(\tilde{\omega}_E)},$$
$$\|v - \mathcal{I}_\mathbb{V} v\|_{L^2(\partial E)} \lesssim h_E^{1/2} \, \|v\|_{H_0^1(\tilde{\omega}_E)}$$

for all $v \in H_0^1(\Omega)$, see also Theorem 3.1.17. Substituting these estimates into estimate (3.14) and applying Cauchy-Schwarz inequality we derive

$$\langle \mathcal{R}, v \rangle \lesssim \Big(\sum_{E \in \mathcal{G}} h_E^2 \, \|R\|_{L^2(E)}^2 \Big)^{\frac{1}{2}} \Big(\sum_{E \in \mathcal{G}} \|v\|_{H_0^1(\tilde{\omega}_E)}^2 \Big)^{\frac{1}{2}}$$
$$+ \Big(\sum_{E \in \mathcal{G}} h_E \, \|j\|_{L^2(\partial E)}^2 \Big)^{\frac{1}{2}} \Big(\sum_{E \in \mathcal{G}} \|v\|_{H_0^1(\tilde{\omega}_E)}^2 \Big)^{\frac{1}{2}}.$$

We then make use of the fact that every patch only contains a bounded number of elements depending on the shape regularity $\sigma_{\mathcal{G}}$. This allows us to estimate $\sum_{E \in \mathcal{G}} \|v\|_{H^1(\tilde{\omega}_E)}^2 \lesssim \|v\|_{H^1(\Omega)}^2$. Additionally, we employ Young's inequality to find

$$\langle \mathcal{R}, v \rangle \lesssim \|v\|_{H_0^1(\Omega)} \Big(\sum_{E \in \mathcal{G}} h_E^2 \, \|R\|_{L^2(E)}^2 + h_E \, \|j\|_{L^2(\partial E)}^2 \Big)^{1/2}.$$

Since the norms $\|\cdot\|_{H_0^1(\Omega)}$ and $\|\cdot\|_\Omega$ are equivalent, we particularly see $\|v\|_{H^1(\Omega)} \leq \|v\|_\Omega / \mathcal{C}_-$ and thus, in light of the equivalence of error and residual from Lemma 3.3.11 we arrive at

$$\|u - U\|_\Omega^2 \leq \|\mathcal{R}\|_*^2 \leq \mathcal{C}_h \sum_{E \in \mathcal{G}} h_E^2 \, \|R\|_{L^2(E)}^2 + h_E \, \|j\|_{L^2(\partial E)}^2, \quad (3.15)$$

where $\mathcal{C}_h$ only depends on the shape regularity $\sigma_{\mathcal{G}}$, the dimension d and $\mathcal{C}_-$. From Lemma 2.1.3 we recall that $\mathcal{C}_-$ only depends on the globally smallest eigenvalue a_- of $\mathbf{A}$, see property (2.3a).

Motivated by above estimate (3.15), we introduce the following local error indicators.

Definition 3.3.12 (Element indicators)
For the element residual $R \in L^2(\Omega)$ and the jump residual $j \in L^2(\Gamma_{\mathcal{G}})$ as defined in (3.11) respectively (3.12), we define for any $E \in \mathcal{G}$ the *element indicator*

$$\mathcal{E}^2(U, E) := h_E^2 \, \|R\|_{L^2(E)}^2 + h_E \, \|j\|_{L^2(\partial E)}^2.$$

Conveniently, we abbreviate $\mathcal{E}^2(U, \mathcal{G}) := \sum_{E \in \mathcal{G}} \mathcal{E}^2(U, E)$.

Employing this notation and summarizing the above, we obtain the well-known main result of this section.

Theorem 3.3.13 (Upper bound for elliptic problems)
*Let $u \in H_0^1(\Omega)$ and $U \in \mathbb{V}$ be the exact solution and the Galerkin
solution respectively. Then the a posteriori error estimate*

$$\|u - U\|_\Omega^2 \leq \|\mathcal{R}\|_*^2 \leq \mathcal{C}_h \mathcal{E}^2(U, \mathcal{G})$$

*is valid where the constant $\mathcal{C}_h$ only depends on the shape coefficient,
the dimension d and the globally smallest eigenvalue a_- of $\mathbf{A}$. In
particular, $\mathcal{C}_h$ is independent of the zero order term c.*

Remark 3.3.14
We will make essential use of this upper bound for elliptic problems in the derivation of an upper bound for parabolic problems in Section 3.4.1. We point out that particularly with respect to parabolic problems, the fact that the involved constant $\mathcal{C}_h$ is independent of the zero order term is very important since in this case, the currently employed time step size enters the zero order term, see Remark 3.3.22.

Lower bound

For completeness, we also give a lower bound. However, since we derive a special lower bound tailored to the specific requirements of time dependent problems in Section 3.4.2, we do not provide a proof of the following well-known lower bound for elliptic problems. Instead, we exemplarily refer to [64, 2, 40] for details and only point out that the required *local* continuity of the bilinear form $\mathcal{B}$ is satisfied in the present case. More precisely, for any subset $\omega \subset \Omega$ we have

$$\mathcal{B}(v, w) \leq \mathcal{C}_{\mathcal{B}} \|v\|_\omega \|w\|_\Omega \qquad \text{for all } w \text{ with } \operatorname{supp} w \subset \overline{\omega},$$

which follows on the same lines as the global continuity given in Lemma 2.1.3.

Theorem 3.3.15 (Local lower bound for elliptic problems)
Each error indicator is bounded by the discretization error up to oscillation, i. e.,

$$\mathcal{E}(U, E) \lesssim \|u - U\|_{\omega_E} + \operatorname{osc}(U, E) \qquad \text{for all } E \in \mathcal{G}.$$

Here, $\operatorname{osc}(U, E)$ denotes a higher order oscillation term of the type

$$\operatorname{osc}(U, E) := h_E \left\| R - \overline{R}_E \right\|_{L^2(\omega_E)} + h_E^{1/2} \left\| j - \overline{j}_S \right\|_{L^2(\partial E \setminus \partial \Omega)},$$

where $\overline{R}_E$ and $\overline{j}_S$ denotes the mean value of R and j on the element E respectively the side $S \in \mathcal{S}_E$. The constant hidden in $\lesssim$ only depends on the shape coefficients σ_E of the simplices in ω_E and the dimension d.

3.3.5 Adaptivity for stationary problems

In this subsection, we present the standard adaptive algorithm to approximate the exact solution u of (3.4). The basic idea behind this is the following: For a given finite element space $\mathbb{V}(\mathcal{G}_0)$, we compute the Galerkin approximation $U_0 \in \mathbb{V}(\mathcal{G}_0)$ of (3.6). We then employ Theorem 3.3.13 to estimate the discretization error $\|u - U\|_\Omega$ and provided this estimate is small enough, we accept the approximation U of u. On the other hand, if the estimated error is too big, we may want to compute a better approximation. Since the only mean to influence the approximation U is the choice of the finite dimensional space, we wish to construct a space $\mathbb{V}(\mathcal{G}_1)$ such that the corresponding Galerkin solution $U_1 \in \mathbb{V}(\mathcal{G}_1)$ approximates u more accurately. As according Céa's Lemma (cf. Lemma 3.3.8) the Galerkin solution U_0 is the quasi best approximation in the current space $\mathbb{V}(\mathcal{G}_0)$, it seems appropriate to create the new space $\mathbb{V}(\mathcal{G}_1)$ by enlarging $\mathbb{V}(\mathcal{G}_0)$. To that end, we recall that the employed finite element spaces are nested in the sense that $\mathbb{V}(\mathcal{G}) \subset \mathbb{V}(\mathcal{G}_+)$ if $\mathcal{G}_+ \geq \mathcal{G}$ is a refinement of $\mathcal{G}$. Hence, in order to reduce the discretization error, we refine the mesh. We emphasize that in light of the upper and lower bound on the discretization error given in Theorems 3.3.13 and 3.3.15, the element indicators introduced in 3.3.12 are closely linked to the *local* discretization error. Thanks to this, we may also use the error indicators to adaptively decide *how* to refine the mesh. Recalling the mesh refinement framework introduced in Section 3.1.1, we point out that this is accomplished by defining the set $\mathcal{M}^m$ of marked elements and corresponding element markers. The standard adaptive method is then characterized by the iteration

$$\text{SOLVE} \longrightarrow \text{ESTIMATE} \longrightarrow \text{MARK} \longrightarrow \text{REFINE.}$$
$$\text{(SEMR)}$$

Starting from an initial triangulation $\mathcal{G}_0$, this iteration generates a sequence of meshes $\{\mathcal{G}_k\}_{k \geq 0}$ and Galerkin solutions $\{U_k\}_{k \geq 0}$ corresponding to the finite element spaces $\{\mathbb{V}(\mathcal{G}_k)\}_{k \geq 0}$. Algorithm 1 implements this iteration in an infinite loop, which is only exited if a given positive tolerance TOL is reached, i. e., $\mathcal{E}^2(U_k, \mathcal{G}_k) \leq$ TOL for some $k \geq 0$.

Algorithm 1 Adaptation for elliptic problems

Initialize $\mathcal{G}_0$ and set $k := 0$;
loop forever
    $U_k := \mathsf{SOLVE}(\mathbb{V}(\mathcal{G}_k))$;
    $\{\mathcal{E}(U_k, E)\}_{E \in \mathcal{G}_k} := \mathsf{ESTIMATE}(U_k, \mathcal{G}_k)$;
    if $\mathcal{E}^2(U_k, \mathcal{G}_k) \leq$ TOL **then**
        break;
    end if
    $\mathcal{M}_k^{\mathrm{m}} := \mathsf{MARK}(\{\mathcal{E}(U_k, E)\}_{E \in \mathcal{G}_k}, \mathcal{G}_k)$;
    $\mathcal{G}_{k+1} := \mathsf{REFINE}(\mathcal{G}_k, \mathcal{M}_k^{\mathrm{m}})$;
    $k := k + 1$;
end loop forever

Whereas we already introduced the module REFINE in Section 3.1.1, we state the properties of the remaining modules of Algorithm 1 in the following.

Properties 3.3.16 (SOLVE)
The module $U := \mathsf{SOLVE}(\mathbb{V})$ computes and returns the Galerkin solution corresponding to the finite element space $\mathbb{V}$, i. e., $U \in \mathbb{V}$ such that

$$\mathcal{B}(U, V) = \langle f, V \rangle \qquad \text{for all } V \in \mathbb{V}.$$

Properties 3.3.17 (ESTIMATE)
The module $\{\mathcal{E}(U, E)\}_{E \in \mathcal{G}} := \mathsf{ESTIMATE}(U, \mathcal{G})$ computes and returns the set of a posteriori error indicators $\{\mathcal{E}(U, E)\}_{E \in \mathcal{G}}$ introduced in Definition 3.3.12.

Properties 3.3.18 (MARK)
The module $\mathcal{M}^{\mathrm{m}} := \mathsf{MARK}(\{\mathcal{E}(U, E)\}_{E \in \mathcal{G}}, \mathcal{G})$ generates a set $\mathcal{M}^{\mathrm{m}} \subset \mathcal{G} \times \mathbb{N}$ of marked elements $\mathcal{M} \subset \mathcal{G}$ and corresponding element markers $m(E)$, cf. Section 3.1.1. For that cause, the module relies on the local error indicators $\{\mathcal{E}(U, E)\}_{E \in \mathcal{G}}$.

The module MARK plays an important role in the adaptive method presented in Algorithm 1 as it transforms information about the error (indicator) into a decision how to refine the mesh.

We state three common strategies for implementing this process. Since all three strategies assign the same element marker to each marked element $E \in \mathcal{M}$ (usually one chooses $m(E) = d$), we neglect the element marker and completely focus on the plain set $\mathcal{M} \subset \mathcal{G}$ of marked elements; $\mathcal{M}^{\mathrm{m}}$ is then implicitly defined as $\mathcal{M}^{\mathrm{m}} := \{(E, 1) \mid E \in \mathcal{M}\}$.

MS The *maximum strategy* selects all elements for refinement whose error indicator exceeds a certain share of the maximum error indicator of the whole mesh. More precisely, for a given parameter $\theta \in [0, 1]$ we set

$$\mathcal{M} := \left\{ E \in \mathcal{G} \mid \mathcal{E}(U, E) \geq \theta\, \mathcal{E}_{\mathcal{G}}^{\mathrm{max}} \right\},$$

where $\mathcal{E}_{\mathcal{G}}^{\mathrm{max}} := \max_{E \in \mathcal{G}} \mathcal{E}(U, E)$ is the largest error indicator on the mesh.

ES The *equidistribution strategy* aims at distributing the error estimate $\mathcal{E}(U, \mathcal{G})$ equally among all elements $E \in \mathcal{G}$ in the sense that all local indicators $\mathcal{E}(U, E)$ are the same size. Assuming this equidistribution, we find

$$\mathcal{E}(U, \mathcal{G}) = \left(\sum_{E \in \mathcal{G}} \mathcal{E}^2(U, E) \right)^{1/2} = \sqrt{\#\mathcal{G}}\, \mathcal{E}(U, E)$$

with $\#\mathcal{G}$ denoting the number of elements in $\mathcal{G}$. This motivates marking all elements with error indicators larger than $\mathcal{E}(U, \mathcal{G})/\sqrt{\#\mathcal{G}}$ for refinement. With a safety parameter $\theta \in (0, 1)$, the equidistribution strategy sets

$$\mathcal{M} := \left\{ E \in \mathcal{G} \mid \mathcal{E}(U, E) \geq \theta\, \frac{\mathcal{E}(U, \mathcal{G})}{\sqrt{\#\mathcal{G}}} \right\}.$$

Note that whereas this version of the equidistribution strategy aims at equidistributing the error estimate $\mathcal{E}(U, \mathcal{G})$, another popular version is geared towards equidistributing the available tolerance TOL and hence marks

$$\tilde{\mathcal{M}} := \left\{ E \in \mathcal{G} \mid \mathcal{E}(U, E) \geq \theta\, \frac{\mathrm{TOL}}{\sqrt{\#\mathcal{G}}} \right\}.$$

DS *Dörfler's strategy* aims at being able to control the overall estimate of the discretization error by just considering the estimated error on the marked elements. More precisely, for a parameter $\theta \in (0, 1]$, the set of marked elements $\mathcal{M} \subset \mathcal{G}$ is chosen such that

$$\mathcal{E}(U, \mathcal{M}) \geq \theta \mathcal{E}(U, \mathcal{G}).$$

For efficiency reasons one desires to mark as few elements as possible to establish above control. This can be accomplished by iteratively growing $\mathcal{M}$ using a greedy algorithm.

We are interested in the termination of Algorithm 1. Apparently, the infinite loop is halted if and only if the **break**-command is hit. Since this in turn is the case if and only if $\mathcal{E}^2(U_k, \mathcal{G}_k) \leq \mathsf{TOL}$ for the given tolerance, the above question of termination is closely related to the convergence $\mathcal{E}^2(U_k, \mathcal{G}_k) \to 0$ for $k \to \infty$.

This convergence is not at all self-evident. In particular, we cannot expect the adaptive algorithm to provide arbitrary small mesh sizes throughout the whole domain. Rather, depending on the structure of the exact solution u, we expect that in general there appear elements $E \in \mathcal{G}_K$ for some $K \geq 0$, which are never refined in the process. As a consequence, the maximal mesh size $H_k := \max_{E \in \mathcal{G}_k} h_E$ of the mesh $\mathcal{G}_k$ will *not* tend to zero as $k \to \infty$ but we rather have $\lim_{k \to \infty} H_k = C > 0$. Such effects especially eliminate the possibility of using an a priori estimate for establishing the convergence $\mathcal{E}^2(U_k, \mathcal{G}_k) \to 0$ as $k \to \infty$.

With this in mind, properties of the modules SOLVE, ESTI-MATE, MARK, and REFINE have to be exploited in order to obtain the desired convergence. Particularly the module MARK is important in this context, as it must guarantee that any error source will eventually be eliminated. For this reason, we ask the additional abstract property of the module MARK following [36].

Properties 3.3.19 (Additional property of MARK)
We assume that given a mesh $\mathcal{G}$ and error indicators $\{\mathcal{E}(U, E)\}_{E \in \mathcal{G}}$, the set $\mathcal{M}$ of marked elements implicitly produced by $\mathcal{M}^{\mathrm{m}} = \mathsf{MARK}(\{\mathcal{E}(U, E)\}_{E \in \mathcal{G}}, \mathcal{G})$ satisfies

$$\max\{\mathcal{E}(U, E) \mid E \in \mathcal{G} \setminus \mathcal{M}\} \leq g(\max\{\mathcal{E}(U, E) \mid E \in \mathcal{M}\})$$

for some fixed function $g : \mathbb{R}_0^+ \to \mathbb{R}_0^+$ which is continuous at 0 with $g(0) = 0$.

Remark 3.3.20
This additional assumption on MARK implies that all error indicators in $\mathcal{G}$ may be controlled by the largest indicator in the set of marked elements. Since the marking strategies (MS), (ES), and (DS) all capture the largest indicator on the whole mesh, they satisfy Assumption 3.3.19 with $g(t) := t$.

With these assumptions, we are in the position to apply the fundamental convergence result for elliptic problems, which was established by Morin, Siebert, and Veeser in [36] and extended by Siebert in [59]. This convergence result also plays an essential role in the convergence of the adaptive algorithm for parabolic problems presented and analyzed in Chapter 4, as it provides the convergence of the elliptic sub-problems.

Theorem 3.3.21 (Convergence and termination of adaptation for elliptic problems)
Let the modules SOLVE, ESTIMATE, and MARK satisfy Assumptions 3.3.16, 3.3.17, 3.3.18, and 3.3.19. Then Algorithm 1 reaches any given tolerance TOL > 0 for the error estimate $\mathcal{E}^2(U_k, \mathcal{G}_k)$ within a finite number of iterations, i. e., there is an iteration index K such that

$$\mathcal{E}^2(U_K, \mathcal{G}_K) \leq \text{TOL}.$$

In particular, Algorithm 1 terminates.

Remark 3.3.22 (Elliptic problems arising in the discretization of parabolic problems)
We point out that in elliptic problems emerging from the discretization of a parabolic problem, the current time step size τ_n enters the zero order term of the elliptic equation in the from $\frac{1}{\tau_n}$, cf. Remark 3.3.14. More precisely, in this situation elliptic problems of the type

$$\mathcal{L}u + \frac{1}{\tau_n}u = f + \frac{1}{\tau_n}U_{n-1}$$

are considered, cf. Section 3.2. Since the time step size τ_n is usually decreased consecutively in any single time step of an adaptive algorithm for solving parabolic problems, we realize that the associated elliptic problem changes and that $\frac{1}{\tau_n}$ is unbounded for $\tau_n \to 0$. Thus, we have to see to it that the time step size τ_n will eventually be fixed and bounded away from zero in any given time step. Only after having this established, a *fixed* elliptic problem is considered and above convergence result is applicable. We will elaborate on this subject in Sections 4.2 and 4.3.

3.4 A posteriori analysis

After this digression considering stationary problems as they arise in the Euler-Galerkin discretization introduced in Section 3.2, we return to the time dependent problem. An important step towards an adaptive method for solving time dependent problems is a thorough a posteriori error analysis. This will ultimately provide a reliable and efficient error estimator as well as local error indicators. We will use the global error estimator to monitor the discretization error and employ the local error indicators to derive a strategy for appropriately reducing the discretization error if necessary.

Before going into the error analysis, we recall both the weak formulation from Definition 2.1.9 and its discretization from Section 3.2.2. The weak formulation is to find $u \in \mathbb{W}(0, T)$ such that

$$\langle u'(t), v \rangle + \mathcal{B}(u(t), v) = \langle f(t), v \rangle \qquad \text{for all } v \in H_0^1(\Omega),$$
$$\text{f.a.e. } t \in (0, T), \qquad \text{(W)}$$
$$u(0) = u_0.$$

Recalling the notation $\mathbb{V}_n = \mathbb{V}(\mathcal{G}_n)$, the full Euler-Galerkin discretization of this problem reads: For given $U_0 \in \mathbb{V}_0$ find $U_n \in \mathbb{V}_n$, $n = 1, \ldots, N$, such that

$$\frac{1}{\tau_n} \langle U_n - U_{n-1}, V \rangle + \mathcal{B}(U_n, V) = \langle f_n, V \rangle \qquad \text{for all } V \in \mathbb{V}_n.$$
$$\text{(EG)}$$

Moreover, we recall that the discrete solutions $U_n \in \mathbb{V}_n$, $n = 1, \ldots, N$, give rise to the linear interpolation

$$U(t) = \frac{t_n - t}{\tau_n} U_{n-1} + \frac{t - t_{n-1}}{\tau_n} U_n \qquad \text{for } t \in (t_{n-1}, t_n].$$

The goal of the analysis presented in the following is to bound the difference between the exact solution u of (W) and the linear interpolation U of the Euler-Galerkin solutions $\{U_n\}_{n=1,\ldots,N}$ of (EG) in a certain norm. We will refer to this difference as *discretization error*. Throughout this section, u denotes the exact solution of (W) whereas U_n, $n = 1, \ldots, N$, denotes the solution of (EG) in the n-th time step. The related linear interpolation is indicated by U.

For reasons of clarity and comprehensibility, we introduce the following *error indicators*:

$$\mathcal{E}_0^2 := \left(1 + 2\mathcal{C}_B^2\right) \|u_0 - U_0\|_{L^2(\Omega)}^2 \tag{3.16a}$$

$$\mathcal{E}_{f;n}^2 := \mathcal{C}_f \frac{1}{\tau_n} \int_{t_{n-1}}^{t_n} \|f - f_n\|_{L^2(\Omega)}^2 \, dt \tag{3.16b}$$

$$\mathcal{E}_{c\tau;n}^2 := \left(3 + 2\mathcal{C}_B^2\right) \mathcal{C}_B^2 \|U_n - U_{n-1}\|_\Omega^2 \tag{3.16c}$$

$$\mathcal{E}_{\mathcal{G};n}^2(E) := \left(9 + 6\mathcal{C}_B^2\right) \mathcal{C}_h \left(h_E^2 \left\| f_n - \frac{1}{\tau_n}(U_n - U_{n-1}) - \mathcal{L}U_n \right\|_{L^2(E)}^2 \right.$$
$$\left. + h_E \|[\![\mathbf{A}\nabla U]\!]\|_{L^2(\partial E \cap \Omega)}^2 \right) \tag{3.16d}$$

In the last indicator, $E \in \mathcal{G}_n$ denotes an arbitrary element from $\mathcal{G}_n$ and $\mathcal{C}_h$ is the constant from the upper bound for the related elliptic problem, cf. Section 3.2.2 and Theorem 3.3.13. We emphasize that $\mathcal{C}_h$ only depends on the shape regularity $\sigma_{\mathcal{G}_n}$ of $\mathcal{G}_n$, the dimension d and the smallest eigenvalue a_- of $\mathbf{A}$ and is particularly independent of the zero order term of the associated elliptic problem, which contains the current time step size, cf. Section 3.2.2 and Corollary 3.4.4. As we will see in the derivation of estimate (3.24) preceding the fully computable upper bound given shortly in Corollary 3.4.7, the constant $\mathcal{C}_f$ in the second indicator only depends on properties of the differential operator and the domain Ω.

The four indicators will be motivated and justified in the following, where we will find that they can be used to establish upper and lower bounds for the discretization error. As we shall see later, the labels *initial*, *consistency*, *coarsen-time* and *space indicator* are appropriate for the individual indicators in (3.16). Conveniently, referring to the spatial indicator, we also define

$$\mathcal{E}_{\mathcal{G};n}^2(\mathcal{M}) := \sum_{E \in \mathcal{M}} \mathcal{E}_{\mathcal{G};n}^2(E)$$

for any set $\mathcal{M} \subset \mathcal{G}_n$.

Similarly to the derivation of an a posteriori estimate for elliptic problems in Section 3.3.4, we now introduce the *time dependent* residual for problem (EG).

Definition 3.4.1 (Residual)
For any $v \in \mathbb{W}(0,T)$, we define the *residual* $\mathcal{R}(v) \in L^2(0,T; H^{-1}(\Omega))$ via

$$\langle \mathcal{R}(v)\,,\,w \rangle_{H^{-1}(\Omega) \times H_0^1(\Omega)} := \langle f\,,\,w \rangle - \langle v'\,,\,w \rangle - \mathcal{B}(v\,,\,w)$$

for all $w \in H_0^1(\Omega)$.

Note that the residual $\mathcal{R}(v)$ is time dependent as it particularly involves the time dependent right hand side f. Recalling that $f \in L^2(0,T; L^2(\Omega))$ only possesses L^2-regularity in time, it is clear that also the residual is at best L^2-regular with respect to time. Thus, the above definition should be read in an "almost everywhere in time" sense which matches perfectly also the regularity of the function $v \in \mathbb{W}(0,T)$. In the following we omit the subscript in dual pairings involving the residual.

By using the weak formulation (W), we may connect the residual to the exact solution $u \in \mathbb{W}(0,T)$ observing

$$\langle \mathcal{R}(v)\,,\,w \rangle = \langle u' - v'\,,\,w \rangle + \mathcal{B}(u - v\,,\,w) \qquad \text{for all } w \in H_0^1(\Omega).$$
(3.17)

Again in analogy to the derivation of the a posteriori error estimate for elliptic problems in Section 3.3.4, we first show the equivalence of error and residual. In a second step, we will then estimate a dual norm of the residual which eventually establishes upper and lower bounds for the discretization error. Error bounds similar to the ones presented in the following were originally derived by Verfürth in [65].

Lemma 3.4.2 (Equivalence of error and residual)
Let $0 \le s_1 \le s_2 \le T$ and assume $v \in \mathbb{W}(0,T)$. Then for any $w \in L^2(0,T; H_0^1(\Omega))$ the lower bound

$$\int_{s_1}^{s_2} \langle \mathcal{R}(v)\,,\,w \rangle \, dt \lesssim \|u - v\|_{\mathbb{W}(s_1,s_2)} \, \|w\|_{L^2(s_1,s_2; H_0^1(\Omega))}$$

holds and the constant hidden in $\lesssim$ only depends on the continuity constant $\mathcal{C}_\mathcal{B}$ of the bilinear form $\mathcal{B}(\cdot\,,\,\cdot)$. Conversely, the error can be bounded from above by

$$\|u - v\|_{\mathbb{W}(0,s_2)}^2 \le (1 + 2\mathcal{C}_\mathcal{B}^2) \, \|(u - v)(0)\|_{L^2(\Omega)}^2$$

$$+ (3 + 2\mathcal{C}_\mathcal{B}^2) \int_0^{s_2} \|\mathcal{R}(v)\|_*^2 \, dt.$$

Proof. First, we conveniently recall for $v \in \mathbb{W}(0,T)$ the definition

$$\|v\|^2_{\mathbb{W}(s_1,s_2)} := \int_{s_1}^{s_2} \|v'\|^2_* + \|v\|^2_\Omega \, dt$$

from Definition 2.1.7. We then start by deriving the lower bound. For this cause, we integrate equation (3.17) from s_1 to s_2 with $0 \leq s_1 \leq s_2 \leq T$. Using the continuity of $\mathcal{B}(\cdot\,,\cdot)$ and Cauchy-Schwarz inequality, we find

$$\int_{s_1}^{s_2} \langle \mathcal{R}(v)\,,\,w\rangle \, dt \leq \int_{s_1}^{s_2} \|u'-v'\|_* \|w\|_\Omega + \mathcal{C}_\mathcal{B} \|u-v\|_\Omega \|w\|_\Omega \, dt$$

$$\leq \left(\int_{s_1}^{s_2} \left(\|u'-v'\|_* + \mathcal{C}_\mathcal{B} \|u-v\|_\Omega \right)^2 dt \right)^{1/2} \|w\|_{L^2(s_1,s_2;H_0^1(\Omega))},$$

which we estimate further by Young's inequality to obtain

$$\leq \left(\max\{2, 2\mathcal{C}_\mathcal{B}^2\} \right)^{1/2} \left(\int_{s_1}^{s_2} \|u'-v'\|^2_* \right.$$

$$\left. + \|u-v\|^2_\Omega \, dt \right)^{1/2} \|w\|_{L^2(s_1,s_2;H_0^1(\Omega))}.$$

With this, the suggested lower bound is established and we proceed with deriving the upper bound. To that end, we first notice

$$\frac{1}{2}\frac{d}{dt} \|u-v\|^2_{L^2(\Omega)} + \|u-v\|^2_\Omega = \langle u'-v'\,,\,u-v\rangle + \mathcal{B}(u-v\,,\,u-v)$$

and using (3.17) as well as Cauchy-Schwarz and Young's inequalities, we estimate

$$= \langle \mathcal{R}(v)\,,\,u-v\rangle \leq \|\mathcal{R}(v)\|_* \|u-v\|_\Omega$$

$$\leq \frac{1}{2} \|\mathcal{R}(v)\|^2_* + \frac{1}{2} \|u-v\|^2_\Omega. \tag{3.18}$$

With this, we established

$$\frac{d}{dt} \|u-v\|^2_{L^2(\Omega)} + \|u-v\|^2_\Omega \leq \|\mathcal{R}(v)\|^2_*$$

and since we are particularly interested in the term $\int_0^{s_2} \|u - v\|_\Omega^2 \, dt$, we integrate this estimate from 0 to s_2 to find

$$\|(u - v)(s_2)\|_{L^2(\Omega)}^2 + \int_0^{s_2} \|u - v\|_\Omega^2 \, dt$$
$$\leq \|(u - v)(0)\|_{L^2(\Omega)}^2 + \int_0^{s_2} \|\mathcal{R}(v)\|_*^2 \, dt. \qquad (3.19)$$

It remains to treat the term $\int_0^{s_2} \|u' - v'\|_*^2 \, dt$ and from equation (3.17) we find

$$\|(u - v)'\|_* \leq \|\mathcal{R}(v)\|_* + C_\mathcal{B} \|u - v\|_\Omega \, .$$

Integrating this estimate and once more applying Young's inequality implies

$$\int_0^{s_2} \|(u - v)'\|_*^2 \, dt \leq \int_0^{s_2} 2 \|\mathcal{R}(v)\|_*^2 + 2C_\mathcal{B}^2 \|u - v\|_\Omega^2 \, dt,$$

where we may estimate the last term using equation (3.19)

$$\leq 2C_\mathcal{B}^2 \|(u - v)(0)\|_{L^2(\Omega)}^2 + (2 + 2C_\mathcal{B}^2) \int_0^{s_2} \|\mathcal{R}(v)\|_*^2 \, dt.$$

Combining this with estimate (3.19) proves the upper bound. □

Remark 3.4.3
With the same techniques as used in the foregoing proof it is also possible to additionally bound the term $\|u - U\|_{L^\infty(0,T;L^2(\Omega))}^2$. This additional bound can be obtained by observing that estimate (3.19) holds for arbitrary $s_2 \in (0, T)$, see also [65] and [14].

3.4.1 Upper bound

In light of Lemma 3.4.2, the residual $\mathcal{R}(v)$ is capable of measuring the difference between the exact solution u and *any* given function $v \in \mathbb{W}(0, T)$. In particular, choosing v as the linear interpolation $U \in \mathbb{W}(0, T)$ of the sequence of solutions $\{U_n\}_{n=1,\ldots,N}$ to (EG), a (dual) norm of the residual estimates the discretization error $\|u - U\|_{\mathbb{W}(0,T)}$. Note that the residual does neither possess nor need any information about the structure of $U \in \mathbb{W}(0,T)$ or how it was computed. Conversely, in order to derive an actually *computable* bound for the dual norm $\|\mathcal{R}(U)\|_*$, we are certainly interested in exploiting as many information as possible. Particularly we want to take into account that in each time step, U_n is the approximate solution of an elliptic problem, hoping to be able to apply some of the results for elliptic problems from Section 3.3. With this motivation, we aim at splitting the residual into three parts reflecting the structure of equation (EG).

From the definition of the residual and the linear interpolation $U(t)$ we find for all $t \in (t_{n-1}, t_n]$.

$$\langle \mathcal{R}(U)(t), w \rangle = \langle f, w \rangle - \langle U'(t), w \rangle - \mathcal{B}(U(t), w)$$
$$= \langle f, w \rangle - \frac{1}{\tau_n}\langle U_n - U_{n-1}, w \rangle - \mathcal{B}(U(t), w).$$

Note that the bilinear form $\mathcal{B}$ has a *time dependent* argument in this representation whereas in (EG), only the *stationary* term $\mathcal{B}(U_n, w)$ is involved. Moreover, we observe that while (EG) only contains the averaged right hand side f_n which is constant with respect to time, the residual involves the full right hand side f in an L^2-sense. Since we want to profit from elliptic theory, the first step towards an appropriate splitting is to enforce the presence of the *elliptic* residual associated to the single equations of (EG); we hence introduce the *spatial residual*

$$\langle \mathcal{R}_h(U)(t), v \rangle := \langle f_n, v \rangle - \frac{1}{\tau_n}\langle U_n - U_{n-1}, v \rangle - \mathcal{B}(U_n, v)$$

for all $t \in (t_{n-1}, t_n]$. Since we want to treat the data error $\langle f - f_n, v \rangle$ explicitly, we further introduce the *temporal residual*

$$\langle \mathcal{R}_\tau(U)(t), v \rangle := \mathcal{B}(U_n - U(t), v) \qquad \text{for all } t \in (t_{n-1}, t_n].$$

With these notions, we may split the full residual into

$$\langle \mathcal{R}(U)(t), v \rangle = \langle \mathcal{R}_h(U)(t), v \rangle + \langle \mathcal{R}_\tau(U)(t), v \rangle + \langle f - f_n, v \rangle \quad (3.20)$$

for all $t \in (t_{n-1}, t_n]$.

In the following, we estimate the three components of the residual individually in order to bound the dual norm $\|\mathcal{R}(U)\|_*$ which is related to the discretization error $\|u - U\|_{\mathbb{W}(0,T)}$ by Lemma 3.4.2. First, we consider the spatial residual. Since it was designed to be the residual of the associated elliptic problem, we essentially benefit from the results from Section 3.3.3 to prove the following corollary.

Corollary 3.4.4 (Estimation of spatial residual)
With the notion of the spatial indicator from (3.16), the following upper bound for the spatial residual holds for all $t \in (t_{n-1}, t_n]$, $n = 1, \ldots, N$:

$$\|\mathcal{R}_h(U)(t)\|_*^2 \leq \frac{1}{9 + 6C_{\mathcal{B}}^2} \, \mathcal{E}_{\mathcal{G};n}^2(\mathcal{G}_n)$$

The constant C_h contained in $\mathcal{E}_{\mathcal{G};n}^2(\mathcal{G}_n)$ is particularly independent of the current time step size τ_n.

Proof. We note that for a fixed $t \in (t_{n-1}, t_n]$, the spatial residual $\langle \mathcal{R}_h(U)(t), v \rangle$ as desired is the residual of the following elliptic problem: Find $u_n \in H_0^1(\Omega)$ such that

$$\mathcal{B}(u_n, v) + \frac{1}{\tau_n}\langle u_n, v \rangle = \langle f_n + \frac{1}{\tau_n}U_{n-1}, v \rangle \qquad \text{for all } v \in H_0^1(\Omega).$$

Since the $\tau_n > 0$ is a given constant and $U_{n-1} \in \mathbb{V}_{n-1} \subset H_0^1(\Omega)$ is also known in the n-th time step, this problem is indeed elliptic and Theorem 3.3.13 directly yields

$$\|\mathcal{R}_h(U)(t)\|_*^2 \leq C_h \sum_{E \in \mathcal{G}_n} \left(h_E^2 \left\| f_n - \frac{1}{\tau_n}(U_n - U_{n-1}) - \mathcal{L}U_n \right\|_{L^2(E)}^2 \right.$$
$$\left. + h_E \left\| [\![\mathbf{A}\nabla U]\!] \right\|_{L^2(\partial E \cap \Omega)}^2 \right)$$
$$= \frac{1}{9 + 6C_{\mathcal{B}}^2} \, \mathcal{E}_{\mathcal{G};n}^2(\mathcal{G}_n).$$

As the constant C_h is independent of the zero order term by Theorem 3.3.13, it is independent of τ_n. $\qquad\square$

We continue with estimating the temporal residual.

Lemma 3.4.5 (Estimation of temporal residual)
For the temporal residual $\mathcal{R}_\tau(U)$ *it holds*

$$\int_{t_{n-1}}^{t_n} \left\| \mathcal{R}_\tau(U)(t) \right\|_*^2 \, dt \leq \mathcal{C}_\mathcal{B}^2 \frac{\tau_n}{3} \left\| U_n - U_{n-1} \right\|_\Omega^2.$$

Proof. From the definition of the temporal residual, it follows for $t \in (t_{n-1}, t_n]$

$$\left\| \mathcal{R}_\tau(U)(t) \right\|_* = \sup_{v \in H_0^1(\Omega)} \frac{\mathcal{B}(U_n - U(t), v)}{\|v\|_\Omega} \leq \mathcal{C}_\mathcal{B} \left\| U_n - U(t) \right\|_\Omega.$$

Together with the equality

$$U_n - U(t) = \frac{t_n - t}{\tau_n} (U_n - U_{n-1}) \qquad \text{for all } t \in (t_{n-1}, t_n],$$

which we find directly from the definition of the linear interpolation $U(t)$, this implies

$$\int_{t_{n-1}}^{t_n} \left\| \mathcal{R}_\tau(U)(t) \right\|_*^2 \, dt \leq \mathcal{C}_\mathcal{B}^2 \left\| U_n - U_{n-1} \right\|_\Omega^2 \int_{t_{n-1}}^{t_n} \left(\frac{t_n - t}{\tau_n} \right)^2 \, dt$$

$$= \mathcal{C}_\mathcal{B}^2 \frac{\tau_n}{3} \left\| U_n - U_{n-1} \right\|_\Omega^2.$$

$\square$

With estimates for both the spatial and the temporal residual at hand, we are now in the position to state an upper bound on the discretization error.

Theorem 3.4.6 (Upper bound)
Let u be the exact solution and let U be the linear interpolation of the sequence of Euler-Galerkin solutions. Then the following upper bound holds for all m with $0 \leq m \leq N$:

$$\|u - U\|_{\mathrm{W}(0,t_m)}^2 \leq \mathcal{E}_0^2 + \sum_{n=1}^{m} \tau_n \left(\mathcal{E}_{\mathcal{G};n}^2(\mathcal{G}_n) + \mathcal{E}_{c\tau;n}^2 \right)$$

$$+ (9 + 6\mathcal{C}_\mathcal{B}^2) \int_0^{t_m} \|f - f_n\|_*^2 \, dt.$$

Proof. Conveniently, we abbreviate in this proof $C_1 := 1 + 2C_{\mathcal{B}}^2$ and $C_2 := 3 + 2C_{\mathcal{B}}^2$. Then, by the equivalence of error and residual from Lemma 3.4.2 and decomposition (3.20) we find

$$\|u - U\|_{\mathrm{W}(0,t_m)}^2 \leq C_1 \|(u - U)(0)\|_{L^2(\Omega)}^2 + C_2 \int_0^{t_m} \|\mathcal{R}(U)\|_*^2 \, dt$$

$$\leq C_1 \|(u - U)(0)\|_{L^2(\Omega)}^2 + C_2 \int_0^{t_m} \left(\|\mathcal{R}_h(U)\|_* + \|\mathcal{R}_\tau(U)\|_* + \|f - f_n\|_* \right)^2 dt$$

and Young's inequality yields

$$\leq C_1 \|(u - U)(0)\|_{L^2(\Omega)}^2$$
$$+ 3C_2 \int_0^{t_m} \|\mathcal{R}_h(U)\|_*^2 \, dt + 3C_2 \int_0^{t_m} \|\mathcal{R}_\tau(U)\|_*^2 \, dt$$
$$+ 3C_2 \int_0^{t_m} \|f - f_n\|_*^2 \, dt. \tag{3.21}$$

We now apply Corollary 3.4.4 to estimate the spatial residual. Since $\mathcal{R}_h(U)$ is constant on each time interval $(t_{n-1}, t_n]$, $n = 1, \ldots, N$, we derive

$$3C_2 \int_0^{t_m} \|\mathcal{R}_h(U)\|_*^2 \, dt \leq \sum_{n=1}^m \tau_n \mathcal{E}_{\mathcal{G};n}^2(\mathcal{G}_n). \tag{3.22}$$

Analogously, we apply Lemma 3.4.5 to estimate the temporal residual on each time interval to find

$$3C_2 \int_0^{t_m} \|\mathcal{R}_\tau(U)\|_*^2 \, dt \leq \sum_{n=1}^m \tau_n \mathcal{E}_{c\tau;n}^2. \tag{3.23}$$

To conclude, we combine (3.21) with the estimates of its individual terms given in (3.22) and (3.23) which yields the assertion. □

In particular, we may choose $m := N$ in the above theorem to obtain an error bound on the whole time interval $(0, T)$. Notice that the upper bound given in Theorem 3.4.6 then contains the term $\int_0^T \|f - f_n\|_*^2 \, dt$ which involves a dual norm of the consistency error $f - f_n$. In order to get an (easily) computable upper bound, it is necessary to further estimate this term. Since both f and f_n have L^2-regularity in space we see

$$\int_0^T \|f - f_n\|_*^2 \, dt = \int_0^T \sup_{v \in H_0^1(\Omega)} \frac{\langle f - f_n , v \rangle_{H^{-1}(\Omega) \times H_0^1(\Omega)}^2}{\|v\|_\Omega^2} \, dt$$

$$= \int_0^T \sup_{v \in H_0^1(\Omega)} \frac{\langle f - f_n , v \rangle_{L^2(\Omega)}^2}{\|v\|_\Omega^2} \, dt,$$

compare also the decomposition of the residual in (3.20). Using Cauchy-Schwarz inequality, we may thus estimate

$$\int_0^T \|f - f_n\|_*^2 \, dt \le \int_0^T \|f - f_n\|_{L^2(\Omega)}^2 \sup_{v \in H_0^1(\Omega)} \frac{\|v\|_{L^2(\Omega)}^2}{\|v\|_\Omega^2} \, dt$$

and since $\|v\|_{L^2(\Omega)} \le C_\mathcal{P} \|v\|_\Omega$ for all $v \in H_0^1(\Omega)$ with Poincaré-Friedrichs constant $C_\mathcal{P}$ (cf. Lemma 2.1.3), we find the computable bound

$$\int_0^T \|f - f_n\|_*^2 \, dt \le C_\mathcal{P}^2 \int_0^T \|f - f_n\|_{L^2(\Omega)}^2 \, dt. \tag{3.24}$$

Setting the constant $C_f := (9 + 6C_\mathcal{B}^2)C_\mathcal{P}^2$, this estimate in combination with Theorem 3.4.6 directly yields the following fully computable upper bound.

Corollary 3.4.7 (Fully computable upper bound)
With the notion of the error indicators introduced in (3.16), the following fully computable upper bound holds:

$$\|u - U\|_{\mathbb{W}(0,T)}^2 \le \mathcal{E}_0^2 + \sum_{n=1}^N \tau_n \left(\mathcal{E}_{\mathcal{G};n}^2(\mathcal{G}_n) + \mathcal{E}_{c\tau;n}^2 + \mathcal{E}_{f;n}^2 \right).$$

Remark 3.4.8 (Alternate estimate for the consistency error)
Opposed to above estimate (3.24), Chen and Feng consider the consistency error differently in the form of $\|f - f_n\|_{L^1(0,T;L^2(\Omega))}^2$ in [14]. In the following, we briefly sketch how this treatment can be achieved. To that end, we recall the proof of the equivalence of

error and residual in Lemma 3.4.2. We notice that the decomposition (3.20) of the residual may already be applied earlier, particularly in equation (3.18). This produces the term $\int_0^T \langle f - f_n , u - U \rangle \, dt$ in the estimation of the discretization error $\|u - U\|_{\mathbb{W}(0,T)}$ which rather treats the consistency error directly instead of hiding it in the residual. Using Hölder's inequality (cf. Theorem 1.2.12), we may then estimate

$$\int_0^T \langle f - f_n , u - U \rangle \, dt \leq \|f - f_n\|_{L^1(0,T;L^2(\Omega))} \, \|u - U\|_{L^\infty(0,T;L^2(\Omega))}$$

and use the continuous embedding $\mathbb{W}(0,T) \hookrightarrow C(0,T;L^2(\Omega))$ for $u - U \in \mathbb{W}(0,T)$ to bound $\|u - U\|_{L^\infty(0,T;L^2(\Omega))} \lesssim \|u - U\|_{\mathbb{W}(0,T)}$. Using this approach, the consistency error then enters the estimate of $\|u - U\|_{\mathbb{W}(0,T)}^2$ in the guise of $\|f - f_n\|_{L^1(0,T;L^2(\Omega))}^2$.

The upper bound on the discretization error stated in Corollary 3.4.7 motivates using the error indicators introduced in (3.16) for error control. However, it remains to justify their titling as "error indicators" which additionally suggests that if the error *indicator* is big this comes with a big actual *error*. We will find this justification in the following subsection.

3.4.2 Lower bound

Beyond justifying the naming of the error indicators from (3.16), the lower bound is of utmost importance for the adaptive algorithm we are aiming for. This is since whereas the upper bound may be employed to check whether a given tolerance on the discretization error is met, a local lower bound may identify levers for actually reducing the discretization error.

We first consider the coarsen-time indicator.

Lemma 3.4.9
For any $n = 1, \ldots, N$, the following estimate holds:

$$\sqrt{\tau_n} \, \|U_n - U_{n-1}\|_\Omega \lesssim \|u - U\|_{\mathbb{W}(t_{n-1}, t_n)} + \sqrt{\tau_n} \, \mathcal{E}_{\mathcal{G};n}(\mathcal{G}_n)$$
$$+ \int_{t_{n-1}}^{t_n} \|f - f_n\|_*^2 \, dt^{1/2}.$$

The constant in $\lesssim$ only depends on the continuity $\mathcal{C}_\mathcal{B}$ of $\mathcal{B}$.

Proof. We conveniently abbreviate $v(t) := U_n - U(t)$ for any given $t \in (t_{n-1}, t_n]$ and as in the proof of Lemma 3.4.5, we find

$$\frac{\tau_n}{3} \|U_n - U_{n-1}\|_\Omega^2 = \|v\|_{L^2(t_{n-1}, t_n; H_0^1(\Omega))}^2. \qquad (3.25)$$

Taking into account the definition of the temporal residual and the splitting (3.20) we further obtain

$$\|v\|_{L^2(t_{n-1}, t_n; H_0^1(\Omega))}^2 = \int_{t_{n-1}}^{t_n} \mathcal{B}(v(t), v(t)) \, dt$$

$$= \int_{t_{n-1}}^{t_n} \langle \mathcal{R}_\tau(U)(t), v(t) \rangle \, dt$$

$$= \int_{t_{n-1}}^{t_n} \langle \mathcal{R}(U)(t), v(t) \rangle \, dt - \int_{t_{n-1}}^{t_n} \langle \mathcal{R}_h(U)(t), v(t) \rangle \, dt$$

$$- \int_{t_{n-1}}^{t_n} \langle f - f_n, v(t) \rangle \, dt. \qquad (3.26)$$

We make use of the equivalence of error and residual and apply the lower bound of Lemma 3.4.2 to the first term yielding

$$\int_{t_{n-1}}^{t_n} \langle \mathcal{R}(U)(t), v(t) \rangle \, dt \lesssim \|u - U\|_{\mathbb{W}(t_{n-1}, t_n)} \|v\|_{L^2(t_{n-1}, t_n; H_0^1(\Omega))},$$

where the constant in $\lesssim$ only depends on the continuity constant $C_\mathcal{B}$ of $\mathcal{B}$. Together with estimating the remaining terms in equation (3.26) by Cauchy-Schwarz inequality, this gives

$$\|v\|_{L^2(t_{n-1}, t_n; H_0^1(\Omega))}^2 \lesssim \|v\|_{L^2(t_{n-1}, t_n; H_0^1(\Omega))} \left(\|u - U\|_{\mathbb{W}(t_{n-1}, t_n)} \right.$$

$$\left. + \int_{t_{n-1}}^{t_n} \|\mathcal{R}_h(U)\|_*^2 \, dt^{1/2} + \int_{t_{n-1}}^{t_n} \|f - f_n\|_*^2 \, dt^{1/2} \right).$$

Dividing both sides by $\|v\|_{L^2(t_{n-1}, t_n; H_0^1(\Omega))}$ and recalling equation (3.25), we deduce

$$\sqrt{\tau_n} \|U_n - U_{n-1}\|_\Omega \lesssim \|u - U\|_{\mathbb{W}(t_{n-1}, t_n)} + \int_{t_{n-1}}^{t_n} \|\mathcal{R}_h(U)\|_*^2 \, dt^{1/2}$$

$$+ \int_{t_{n-1}}^{t_n} \|f - f_n\|_*^2 \, dt^{1/2}.$$

Finally, we apply Corollary 3.4.4 to estimate the spatial residual and since the spatial indicator is constant with respect to time, this reveals

$$\lesssim \|u - U\|_{\mathbb{W}(t_{n-1}, t_n)} + \sqrt{\tau_n}\, \mathcal{E}_{\mathcal{G};n}(\mathcal{G}_n) + \int_{t_{n-1}}^{t_n} \|f - f_n\|_*^2 \, dt^{1/2}.$$

$\square$

We now turn to the spatial estimator. In contrast to the lower bound for elliptic problems cited in Section 3.3.4 where the local error indicator was bounded in terms of the local discretization error (and some oscillation), we now state a more abstract bound in terms of the residual. Like in the upper bound of Section 3.4.1, this presentation also follows the one given by Verfürth in [65], however, there are two differences:

1. Whereas in [65] only the plain heat equation with elliptic operator $\mathcal{L} = -\Delta$ is considered, we employ a more general elliptic operator $\mathcal{L}$, which particularly involves non-constant coefficients and hence, an additional oscillation term arises.

2. Moreover, we point out that in [65] the overlay $\mathcal{G}_{n-1}^n$ of the two meshes $\mathcal{G}_{n-1}$ and $\mathcal{G}_n$ is employed. This is necessary when considering general θ-schemes for temporal discretization (which is done in [65]) like e.g., the Crank-Nicholson scheme which corresponds to $\theta = 1/2$. Such a scheme then involves the term $\mathcal{B}(\theta U_n + (1 - \theta)U_{n-1}, V)$ containing finite element functions from both $\mathbb{V}(\mathcal{G}_{n-1})$ and $\mathbb{V}(\mathcal{G}_n)$ if $0 < \theta < 1$. In this case, particularly the upper bound asks for employing the overlay $\mathcal{G}_{n-1}^n$, as the affine combination $\theta U_n + (1-\theta)U_{n-1}$ is a piecewise polynomial only on this overlay and the corresponding jump residual is only defined on its skeleton. With this technique, however, one has to provide some so-called *transition condition* locally relating the two meshes $\mathcal{G}_{n-1}$ and $\mathcal{G}_n$. The transition condition basically demands that the coarsening used between $\mathcal{G}_{n-1}$ and $\mathcal{G}_n$ is not too harsh. In view of a lower bound, considering the overlay $\mathcal{G}_{n-1}^n$ makes it possible that no oscillation originates from U_{n-1} even though $U_{n-1} \notin \mathbb{V}_n$, but instead, the involved constant depends on the transition condition.

However, since we only consider the implicit Euler method —
which corresponds to $\theta = 1$ and thus does not require consid-
ering the overlay $\mathcal{G}_{n-1}^n$ — we stick to exclusively using $\mathcal{G}_n$ and
treat the previous time step's solution U_{n-1} purely as data.
Consequently, when employing coarsening between $\mathcal{G}_{n-1}$ and
$\mathcal{G}_n$, additional oscillation is introduced (see Lemma 3.4.11),
which suggests also in this case only to use a moderate coarsening to avoid overestimating the error.

For the derivation of the lower bound, we use the well-known
bubble-functions which were introduced by Verfürth, cf. [63, 64].
For an element $E \in \mathcal{G}$, the *element bubble function* $\eta_E \in W_0^{1,\infty}(E)$
is given by

$$\eta_E(x) := (d+1)^{d+1} \prod_{i=0}^{d} \lambda_i^E(x) \qquad \text{for all } x \in E,$$

where λ_i^E, $i = 0, \ldots, d$, are the barycentric coordinates on E, see
Definition 3.1.1. This implies $\operatorname{supp} \eta_E \subset E$ and hence, the supports
of the bubble functions η_E, $E \in \mathcal{G}$, are mutually disjoint. For a
side $S \in \mathcal{S}_\mathcal{G}$, the *side* or *edge bubble function* $\eta_S \in W_0^{1,\infty}(\omega_S)$ is
defined on the patch ω_S by

$$\eta_{S|E}(x) := d^d \prod_{i=0}^{d} \lambda_i^E(x) \qquad \text{for all } E \in \omega_S,\, x \in E.$$

This implies $\operatorname{supp} \eta_S \subset \omega_S$ and particularly, $\operatorname{supp} \eta_S$ intersects
the support of at most two element bubble functions. Vice versa,
$\operatorname{supp} \eta_E$ intersects the support of $d + 1$ different edge bubble functions. Further, we realize that $\operatorname{supp} \eta_S$ intersects the support of at
most $2d$ other edge bubble functions. From [64, Chapter 3] we cite
the following properties of element and edge bubble functions.

Lemma 3.4.10 (Properties of bubble functions)
*Let $E \in \mathcal{G}$ and $S \in \mathcal{S}_\mathcal{G}$. Then for $p_E \in \mathbb{P}_0(E)$ the element bubble
functions satisfy*

$$\langle \eta_E p_E, p_E \rangle_{L^2(E)} \gtrsim \| p_E \|_{L^2(E)}^2, \qquad (3.27a)$$

$$h_E \| \eta_E p_E \|_E \lesssim \| p_E \|_{L^2(E)}. \qquad (3.27b)$$

Moreover, for $p_S \in \mathbb{P}_0(S)$ the edge bubble functions satisfy

$$\langle \eta_S p_S \,, p_S \rangle \gtrsim \|p_S\|_{L^2(S)}^2 \,, \tag{3.28a}$$

$$h_S^{1/2} \, \|\eta_S p_S\|_{\omega_S} \lesssim \|p_S\|_{L^2(S)} \,, \tag{3.28b}$$

$$h_S^{-1/2} \, \|\eta_S p_S\|_{L^2(\omega_S)} \lesssim \|p_S\|_{L^2(S)} \,. \tag{3.28c}$$

The constants hidden in $\lesssim$ depend on the shape coefficient $\sigma_{\mathcal{G}}$, the constants involved in estimating energy norms additionally depend on $\mathcal{C}_+$. The constants hidden in $\gtrsim$ only depend on the bubble functions.

In the following, we consider an arbitrary but fixed time step n. Conveniently, we abbreviate the element and jump residual of the associated elliptic problem in the n-th time step by (cf. Section 3.3.4)

$$R^n := f_n - \mathbf{b} \cdot \nabla U_n - c\, U_n - \frac{1}{\tau_n}(U_n - U_{n-1}) \quad \text{and}$$

$$j^n := [\![\mathbf{A}\nabla U_n]\!]\,,$$

and we denote by $\overline{R}_E^n, \overline{j}_S^n \in \mathbb{R}$ the mean value of R^n and j^n on the element E respectively side S in the sense that $\left\|\overline{R}_E^n\right\|_{L^2(E)} = \|R^n\|_{L^2(E)}$ and $\left\|\overline{j}_S^n\right\|_{L^2(S)} = \|j^n\|_{L^2(S)}$. We explicitly point out that here, U_{n-1} is purely treated as data. Moreover, we define for $v \in H_0^1(\Omega)$ and $t \in (t_{n-1}, t_n]$

$$\langle \overline{\mathcal{R}}_h(t) \,, v \rangle := \sum_{E \in \mathcal{G}_n} \langle \overline{R}_E^n \,, v \rangle_{L^2(E)} + \sum_{S \in \mathcal{S}_{\mathcal{G}_n}} \langle \overline{j}_S^n \,, v \rangle_{L^2(S)}. \tag{3.29}$$

To improve readability, we suppress the superscript n in $\overline{R}_E^n$ and $\overline{j}_S^n$ in the following, since we only consider one particular time step.

Lemma 3.4.11 (Lower bound for spatial residual)
There are functions $w_n \in H_0^1(\Omega)$, $1 \leq n \leq N$, such that the spatial residual satisfies for all $t \in (t_{n-1}, t_n]$

$$\langle \mathcal{R}_h(U)(t) \,, w_n \rangle \geq \mathcal{E}_{\mathcal{G};n}^2(\mathcal{G}_n) + \langle \mathcal{R}_h(U)(t) - \overline{\mathcal{R}}_h(t) \,, w_n \rangle,$$

$$\|w_n\|_\Omega \lesssim \mathcal{E}_{\mathcal{G};n}(\mathcal{G}_n).$$

The constant hidden in $\lesssim$ only depends on the shape regularity of the triangulation, the dimension d as well as the continuity constant $C_\mathcal{B}$ of $\mathcal{B}$ and C_+.

Proof. With the element bubble function η_E and the edge bubble function η_S we define $w_n \in H_0^1(\Omega)$ as

$$w_n := \alpha \sum_{E \in \mathcal{G}_n} h_E^2 \overline{R}_E \eta_E + \beta \sum_{S \in \mathcal{S}_{\mathcal{G}_n}} h_S \overline{j}_S \eta_S$$

with parameters $\alpha, \beta > 0$ to be determined later. First, we establish the bound for the energy norm of w_n. To that end, we exploit that the supports of the element bubble functions are mutually disjoint. The bilinearity of $\mathcal{B}(\cdot, \cdot)$ thus yields

$$
\begin{aligned}
\|w_n\|_\Omega^2 &= \mathcal{B}(w_n, w_n) \\
&= \alpha^2 \sum_{E \in \mathcal{G}_n} h_E^4 \left\| \overline{R}_E \eta_E \right\|_E^2 \\
&\quad + 2\alpha\beta \sum_{E \in \mathcal{G}_n} \sum_{S \in \mathcal{S}_{\mathcal{G}_n}} h_E^2 h_S \mathcal{B}(\overline{R}_E \eta_E, \overline{j}_S \eta_S) \\
&\quad + \beta^2 \sum_{S \in \mathcal{S}_{\mathcal{G}_n}} \sum_{S' \in \mathcal{S}_{\mathcal{G}_n}} h_S h_{S'} \mathcal{B}(\overline{j}_S \eta_S, \overline{j}_S \eta_{S'}).
\end{aligned}
\tag{3.30}
$$

Since $\overline{R}_E$ and $\overline{j}_S$ are constant on each element, we may exploit properties (3.27) and (3.28) of the element and edge bubble functions from Lemma 3.4.10 to bound those terms. For the first term on the right hand side of (3.30), we derive by property (3.27b)

$$h_E^4 \left\| \eta_E \overline{R}_E \right\|_E^2 \lesssim h_E^2 \left\| \overline{R}_E \right\|_{L^2(E)}^2. \tag{3.31}$$

Next, we focus on the last term of equation (3.30). Since the support of the edge bubble function η_S is limited to the corresponding patch ω_S, we have by continuity of $\mathcal{B}$ for "neighboring sides" S and S' in the sense that $\omega_S \cap \omega_{S'} \neq \emptyset$

$$h_S h_{S'} \mathcal{B}(\eta_S \overline{j}_S, \eta_{S'} \overline{j}_S) \leq C_\mathcal{B} h_S h_{S'} \left\| \eta_S \overline{j}_S \right\|_{\omega_S} \left\| \eta_{S'} \overline{j}_S \right\|_{\omega_{S'}}.$$

For non-neighboring sides (in the above sense), this term vanishes. As each side only possesses $2d$ such "neighboring sides", we may thus estimate

$$\sum_{S \in \mathcal{S}_{\mathcal{G}_n}} \sum_{S' \in \mathcal{S}_{\mathcal{G}_n}} h_S h_{S'} \mathcal{B}(\eta_S \bar{\jmath}_S, \eta_{S'} \bar{\jmath}_S) \lesssim \sum_{S \in \mathcal{S}_{\mathcal{G}_n}} h_S^2 \left\| \left| \eta_S \bar{\jmath}_S \right| \right\|_{\omega_S}^2.$$

Before exploiting property (3.28b), we switch our point of view from sides to elements. To that end, we use $h_S \lesssim h_E$ for S being a side of E — with constant depending on the shape regularity — as well as the fact that every element is only contained in $d+1$ patches ω_S:

$$\lesssim \sum_{E \in \mathcal{G}_n} h_E^2 \left\| \left| \bar{\jmath}_S \eta_S \right| \right\|_E^2 \lesssim \sum_{E \in \mathcal{G}_n} h_E \left\| \bar{\jmath}_S \right\|_{L^2(\partial E \cap \Omega)}^2. \quad (3.32)$$

Finally, we consider the remaining second term of the right hand side of equation (3.30). We recall that the support of a given element bubble function only intersects the support of $d+1$ edge bubble functions and analogously to the above estimations, we exploit the properties of the element and edge bubble functions to find

$$\sum_{E \in \mathcal{G}_n} \sum_{S \in \mathcal{S}_{\mathcal{G}_n}} h_E^2 h_S \mathcal{B}(\eta_E \overline{R}_E, \eta_S \bar{\jmath}_S)$$

$$\lesssim \sum_{E \in \mathcal{G}_n} h_E \left\| \overline{R}_E \right\|_{L^2(E)} h_E^{1/2} \left\| \bar{\jmath}_S \right\|_{L^2(\partial E \cap \Omega)}. \quad (3.33)$$

We notice that while the estimates of the first and third term (given in (3.31) and (3.32)) either contain the element *or* the jump residual, estimate (3.33) of the mixed term in equation (3.30) involves both quantities. To get rid of this coupling, we split the element and jump terms via Young's inequality. Substituting this and (3.31), (3.32) into the original equation (3.30) establishes the claimed bound on the energy norm of w_n:

$$\left\| w_n \right\|_\Omega^2 \lesssim \sum_{E \in \mathcal{G}_n} h_E^2 \left\| \overline{R}_E \right\|_{L^2(E)}^2 + h_E \left\| \bar{\jmath}_S \right\|_{L^2(\partial E \cap \Omega)}^2 = \mathcal{E}_{\mathcal{G};n}^2(\mathcal{G}_n)$$

Here, we additionally hide in $\lesssim$ the constants $\alpha, \beta > 0$ which will be determined at the end of this proof and we used the definition of $\overline{R}_E$ and $\bar{\jmath}_S$ as mean values.

We proceed with proving the bound on the residual. For this purpose, let $t \in (t_{n-1}, t_n]$. Using $\overline{\mathcal{R}}_h$ defined in (3.29), we notice that the spatial residual may be represented as

$$\langle \mathcal{R}_h(U)(t) , w_n \rangle = \langle \overline{\mathcal{R}}_h(t) , w_n \rangle + \langle \mathcal{R}_h(U)(t) - \overline{\mathcal{R}}_h(t) , w_n \rangle$$

and we focus on estimating $\langle \overline{\mathcal{R}}_h(t) , w_n \rangle$ in the following. As this eventually asks for defining the parameters α and β appropriately, we use explicit constants $C_1, \dots, C_5$ (without specifying them). Since the element bubble functions η_E vanish on the sides $S \in \mathcal{S}_\mathcal{G}$, we have by definition of $\overline{\mathcal{R}}_h$ and w_n

$$\langle \overline{\mathcal{R}}_h(t) , w_n \rangle = \alpha \sum_{E \in \mathcal{G}_n} h_E^2 \langle \overline{R}_E , \overline{R}_E \eta_E \rangle_{L^2(E)}$$
$$+ \beta \sum_{S \in \mathcal{S}_{\mathcal{G}_n}} h_S \langle \overline{j}_S , \overline{j}_S \eta_S \rangle_{L^2(S)}$$
$$+ \beta \sum_{E \in \mathcal{G}_n} \sum_{S \in \mathcal{S}_\mathcal{G}} h_S \langle \overline{R}_E , \overline{j}_S \eta_S \rangle_{L^2(E)}. \tag{3.34}$$

In the following, we consider the terms of the right hand side of equation (3.34) individually. Starting with the first term, we estimate using property (3.27a)

$$\alpha \sum_{E \in \mathcal{G}_n} h_E^2 \langle \overline{R}_E , \overline{R}_E \eta_E \rangle_{L^2(E)} \geq C_1 \alpha \sum_{E \in \mathcal{G}_n} h_E^2 \left\| \overline{R}_E \right\|_{L^2(E)}^2 \tag{3.35}$$

for some constant $C_1 > 0$. Analogously, we get for the second term using property (3.28a) of the edge bubble functions

$$\beta \sum_{S \in \mathcal{S}_{\mathcal{G}_n}} h_S \langle \overline{j}_S , \overline{j}_S \eta_S \rangle_{L^2(S)} \geq C_2 \beta \sum_{S \in \mathcal{S}_{\mathcal{G}_n}} h_S \left\| \overline{j}_S \right\|_{L^2(S)}^2$$

for some constant $C_2 > 0$. Switching our point of view from sides to elements, we obtain

$$\geq C_3 \beta \sum_{E \in \mathcal{G}_n} h_E \left\| \overline{j}_S \right\|_{L^2(\partial E \cap \Omega)}^2 , \tag{3.36}$$

where we used that $h_S \gtrsim h_E$ for $E \in \omega_S$ with a constant depending on the shape regularity as well as the fact that each interior side is shared by only two elements. Before estimating the last term of the right hand side of (3.34), we first switch our point of view from elements to sides to find

$$\beta \sum_{E \in \mathcal{G}_n} \sum_{S \in \mathcal{S}_{\mathcal{G}_n}} h_S \langle \overline{R}_E \, , \, \eta_S \overline{j}_S \rangle_{L^2(E)}$$

$$= \beta \sum_{S \in \mathcal{S}_{\mathcal{G}_n}} \sum_{E \in \omega_S} h_S \langle \overline{R}_E \, , \, \eta_S \overline{j}_S \rangle_{L^2(E)}.$$

Applying Cauchy–Schwarz inequality before exploiting property (3.28c) of the edge bubble functions we estimate this by

$$\geq -\beta \sum_{S \in \mathcal{S}_{\mathcal{G}_n}} \sum_{E \in \omega_S} h_S \left\| \overline{R}_E \right\|_{L^2(E)} \left\| \eta_S \overline{j}_S \right\|_{L^2(E)}$$

$$\geq -C_4 \beta \sum_{S \in \mathcal{S}_{\mathcal{G}_n}} \sum_{E \in \omega_S} h_S \left\| \overline{R}_E \right\|_{L^2(E)} h_E^{1/2} \left\| \overline{j}_S \right\|_{L^2(S)}.$$

Considering $h_S \lesssim h_E$ for $E \in \omega_S$ with a constant depending on the shape regularity, and observing that an element is only contained in $d + 1$ patches ω_S, we switch our point of view back to elements and estimate further

$$\geq -C_5 \beta \sum_{E \in \mathcal{G}_n} h_E \left\| \overline{R}_E \right\|_{L^2(E)} h_E^{1/2} \left\| \overline{j}_S \right\|_{L^2(\partial E \cap \Omega)}$$

Using Young's inequality, we split

$$\geq -C_5 \beta \sum_{E \in \mathcal{G}_n} \frac{C_5}{2C_3} h_E^2 \left\| \overline{R}_E \right\|_{L^2(E)}^2 + \frac{C_3}{2C_5} h_E \left\| \overline{j}_S \right\|_{L^2(\partial E \cap \Omega)}^2.$$

Substituting this estimate for the last term of the right hand side of equation (3.34) and the estimates (3.35) and (3.36) back into (3.34), we obtain

$$\langle \overline{\mathcal{R}}_h(t) \, , \, w_n \rangle \geq \left(C_1 \alpha - \frac{C_5^2}{2C_3} \beta \right) \sum_{E \in \mathcal{G}_n} h_E^2 \left\| \overline{R}_E \right\|_{L^2(E)}^2$$

$$+ \frac{C_3}{2} \beta \sum_{E \in \mathcal{G}_n} h_E \left\| \overline{j}_S \right\|_{L^2(\partial E \cap \Omega)}^2.$$

Choosing $\beta := 2/C_3 > 0$ and $\alpha := \frac{1}{C_1}(1 + \frac{C_5^2}{C_3^2}) > 0$ and using the definition of $\overline{R}_E$ and $\overline{j}_S$ as mean values completes the proof. □

We make essential use of the previous lemma for proving the following lower bound involving the spatial indicator.

Lemma 3.4.12
The following lower bound, which only involves the spatial estimator, is valid:

$$\sqrt{\tau_n}\mathcal{E}_{\mathcal{G};n}(\mathcal{G}_n) \lesssim \|u - U\|_{\mathbb{W}(t_{n-1},t_n)}$$

$$+ \left(\int_{t_{n-1}}^{t_n} \left\|\left\| \overline{\mathcal{R}}_h - \mathcal{R}_h(U)(t) \right\|\right\|_*^2 + \|f - f_n\|_*^2 \, dt \right)^{1/2}.$$

The constant hidden in $\lesssim$ only depends on the shape regularity of the triangulation, the dimension d as well as the continuity $C_\mathcal{B}$ of $\mathcal{B}$ and C_+.

Proof. We use the bubble functions w_n from Lemma 3.4.11 and recall that for $t \in (t_{n-1}, t_n]$

$$\mathcal{E}_{\mathcal{G};n}^2(\mathcal{G}_n) \leq \langle \mathcal{R}_h(U)(t), w_n \rangle + \langle \overline{\mathcal{R}}_h(t) - \mathcal{R}_h(U)(t), w_n \rangle$$

$$= \langle \mathcal{R}(U)(t), w_n \rangle - \langle \mathcal{R}_\tau(U)(t), w_n \rangle - \langle f - f_n, w_n \rangle$$

$$+ \langle \overline{\mathcal{R}}_h(t) - \mathcal{R}_h(U)(t), w_n \rangle.$$

Moreover, we let $\alpha \geq 0$ be an arbitrary parameter and define $a(t) := (\alpha + 1)\left(\frac{t - t_{n-1}}{\tau_n}\right)^\alpha$. Observing $\int_{t_{n-1}}^{t_n} a(t)dt = \tau_n$ and emphasizing that the spatial indicator $\mathcal{E}_{\mathcal{G};n}^2(\mathcal{G}_n)$ is temporally constant on each time interval, we multiply above equation with $a(t)$ and integrate from t_{n-1} to t_n to find

$$\tau_n \mathcal{E}_{\mathcal{G};n}^2(\mathcal{G}_n) \leq \int_{t_{n-1}}^{t_n} \langle \mathcal{R}(U)(t), a(t)w_n \rangle - \langle \mathcal{R}_\tau(U)(t), a(t)w_n \rangle$$

$$+ \langle \overline{\mathcal{R}}_h(t) - \mathcal{R}_h(U)(t), a(t)w_n \rangle - \langle f - f_n, a(t)w_n \rangle \, dt.$$
$$\text{(3.37)}$$

In the rest of the proof, we estimate the terms of the right hand side individually. Starting with the last two terms, we estimate by

Cauchy-Schwarz and Young's inequality

$$\int_{t_{n-1}}^{t_n} \langle \overline{\mathcal{R}}_h(t) - \mathcal{R}_h(U)(t) , a(t)w_n \rangle - \langle f - f_n , a(t)w_n \rangle \, dt$$

$$\leq \|a\, w_n\|_{L^2(t_{n-1},t_n;H_0^1(\Omega))} \left(2 \int_{t_{n-1}}^{t_n} \left\|\left| \overline{\mathcal{R}}_h(t) - \mathcal{R}_h(U)(t) \right|\right\|_*^2 \right.$$

$$\left. + \|f - f_n\|_*^2 \, dt \right)^{1/2} . \quad (3.38)$$

Before proceeding with estimating the remaining terms from (3.37), we emphasize that w_n is temporally constant and observe

$$\|a\, w_n\|_{L^2(t_{n-1},t_n;H_0^1(\Omega))} = \|w_n\|_\Omega \int_{t_{n-1}}^{t_n} a(t)^2 \, dt^{1/2}$$

$$= \tau_n^{1/2} \frac{\alpha+1}{\sqrt{2\alpha+1}} \, \|w_n\|_\Omega \leq \tau_n^{1/2} \sqrt{2\alpha+1} \, \|w_n\|_\Omega . \quad (3.39)$$

We now focus on the remaining terms from (3.37). From the definition of the temporal residual as well as the representation of the linear interpolation we see for the second term of the right hand side of (3.37)

$$\int_{t_{n-1}}^{t_n} \langle \mathcal{R}_\tau(U)(t) , a(t)w_n \rangle \, dt = \int_{t_{n-1}}^{t_n} a(t) \frac{t_n - t}{\tau_n} \mathcal{B}(U_n - U_{n-1} , w_n) dt$$

$$= \mathcal{B}(U_n - U_{n-1} , w_n) \int_{t_{n-1}}^{t_n} (\alpha+1) \left(\frac{t - t_{n-1}}{\tau_n} \right)^\alpha \frac{t_n - t}{\tau_n} \, dt.$$

To conclude the estimation of this term, we evaluate the integral and use the continuity of $\mathcal{B}$ to find

$$= \tau_n \left(1 - \frac{\alpha+1}{\alpha+2} \right) \mathcal{B}(U_n - U_{n-1} , w_n)$$

$$\lesssim \left(1 - \frac{\alpha+1}{\alpha+2} \right) \tau_n^{1/2} \|U_n - U_{n-1}\|_\Omega \, \tau_n^{1/2} \|w_n\|_\Omega . \quad (3.40)$$

Finally, we estimate the first term of (3.37) by employing the equivalence of error and residual from Lemma 3.4.2 to receive

$$\int_{t_{n-1}}^{t_n} \langle \mathcal{R}(U)(t) \,,\, a(t)w_n \rangle \, dt$$

$$\lesssim \|u - U\|_{\mathbb{W}(t_{n-1}, t_n)} \, \|a \, w_n\|_{L^2(t_{n-1}, t_n, H_0^1(\Omega))} \,, \qquad (3.41)$$

and we recall that we already estimated the norm of $a(t)w_n$ in inequality (3.39).

Now that we estimated all terms of the right hand side of (3.37) individually, we substitute those estimates (3.38), (3.39), (3.40) and (3.41) back into (3.37) to see

$$\tau_n \mathcal{E}_{\mathcal{G};n}^2(\mathcal{G}_n) \lesssim \tau_n^{1/2} \, \|w_n\|_\Omega \left(\sqrt{2\alpha + 1} \, \|u - U\|_{\mathbb{W}(t_{n-1}, t_n)} \right.$$

$$+ \left(1 - \frac{\alpha + 1}{\alpha + 2}\right)\tau_n^{1/2} \, \|U_n - U_{n-1}\|_\Omega$$

$$+ \sqrt{4\alpha + 2} \int_{t_{n-1}}^{t_n} \left\|\!\left\|\overline{\mathcal{R}}_h(t) - \mathcal{R}_h(U)(t)\right\|\!\right\|_*^2$$

$$\left. + \left\|\!\left\|f - f_n\right\|\!\right\|_*^2 \, dt^{1/2} \right).$$

Moreover, we apply Lemma 3.4.9 to further estimate the term $\tau_n^{1/2} \, \|U_n - U_{n-1}\|_\Omega$ and essentially, we use Lemma 3.4.11 to bound $\|w_n\|_\Omega \lesssim \mathcal{E}_{\mathcal{G};n}(\mathcal{G}_n)$. Dividing by $\tau_n^{1/2} \mathcal{E}_{\mathcal{G};n}(\mathcal{G}_n)$ then yields

$$\tau_n^{1/2} \mathcal{E}_{\mathcal{G};n}(\mathcal{G}_n) \lesssim \left(\left(1 - \frac{\alpha + 1}{\alpha + 2}\right) + \sqrt{2\alpha + 1}\right) \|u - U\|_{\mathbb{W}(t_{n-1}, t_n)}$$

$$+ \left(1 - \frac{\alpha + 1}{\alpha + 2}\right)\tau_n^{1/2} \mathcal{E}_{\mathcal{G};n}(\mathcal{G}_n)$$

$$+ \sqrt{4\alpha + 2} \int_{t_{n-1}}^{t_n} \left\|\!\left\|\overline{\mathcal{R}}_h(t) - \mathcal{R}_h(U)\right\|\!\right\|_*^2$$

$$+ \left(2 - \frac{\alpha + 1}{\alpha + 2}\right) \left\|\!\left\|f - f_n\right\|\!\right\|_*^2 \, dt^{1/2}.$$

Notice that the application of Lemma 3.4.9 gives rise to another term $\tau_n^{1/2} \mathcal{E}_{\mathcal{G};n}(\mathcal{G}_n)$ (second term on the right hand side). However, since this term contains a factor $1 - \frac{\alpha + 1}{\alpha + 2}$ which tends to zero as α increases, we may choose α large enough — depending on the constant in $\lesssim$ — such that the term $(1 - \frac{\alpha+1}{\alpha+2})\tau_n^{1/2}\mathcal{E}_{\mathcal{G};n}(\mathcal{G}_n)$ can be absorbed in the left hand side. With this technique, we finally derive

$$\tau_n^{1/2}\mathcal{E}_{\mathcal{G};n}(\mathcal{G}_n) \lesssim \|u - U\|_{W(t_{n-1},t_n)}$$
$$+ \int_{t_{n-1}}^{t_n} \left\|\left|\overline{\mathcal{R}}_h(t) - \mathcal{R}_h(U)(t)\right|\right\|_*^2 + \|f - f_n\|_*^2 \, dt^{1/2}.$$

$\square$

In order to derive a lower bound, we estimate the oscillation term $\left\|\left|\overline{\mathcal{R}}_h(t) - \mathcal{R}_h(U)(t)\right|\right\|_*^2$ with the same techniques as in the derivation of the upper bound for elliptic problems in Section 3.3.4 to find for $t \in (t_{n-1}, t_n]$

$$\left\|\left|\overline{\mathcal{R}}_h(t) - \mathcal{R}_h(U)(t)\right|\right\|_*^2$$
$$\lesssim \sum_{E \in \mathcal{G}_n} h_E^2 \left\|\overline{R}_E - R\right\|_{L^2(E)}^2 + h_E \left\|\overline{j}_S - j\right\|_{L^2(\partial E)}^2$$
$$=: \operatorname{osc}(U_{n-1}, U_n, \mathcal{G}_n).$$

Note that $\operatorname{osc}(U_{n-1}, U_n, \mathcal{G}_n)$ depends on U_{n-1} and U_n since the element and jump residuals involve both U_{n-1} and U_n and the co-efficients $\mathbf{A}$ and $\mathbf{b}$ of the differential operator are generally not constant. Moreover, we point out that $\operatorname{osc}(U_{n-1}, U_n, \mathcal{G}_n)$ is constant on any single time interval. Combining this with Lemma 3.4.9 and Lemma 3.4.12 yields the following lower bound.

Theorem 3.4.13 (Lower bound)
With space and coarsen-time indicators as given in (3.16), the following local (with respect to time) lower bound holds:

$$\tau_n \left(\mathcal{E}_{\mathcal{G};n}^2(\mathcal{G}_n) + \mathcal{E}_{c\tau;n}^2\right) \lesssim \|u - U\|_{W(t_{n-1},t_n)}^2 + \int_{t_{n-1}}^{t_n} \|f - f_n\|_*^2 \, dt$$
$$+ \tau_n \operatorname{osc}(U_{n-1}, U_n, \mathcal{G}_n).$$

Remark 3.4.14
Opposed to the lower bound introduced by Verfürth in [65], the estimate presented above does not contain an L^∞-term of type $\|u - U\|_{L^\infty(t_{n-1},t_n;L^2(\Omega))}^2$ on the right hand side. However, the

proofs are similar to the ones given in [65] and in this sense, an additional L^∞-term on the right hand side is unnecessary. Beyond that, the lower bound stated in Theorem 3.4.13 is summable with respect to time, cf. Corollary 3.4.15 below.

To conclude this section, we summarize this lower bound and the upper bound from Theorem 3.4.6.

Corollary 3.4.15 (Reliability and efficiency)
For any $0 \le m \le N$ we have the upper and lower bound

$$\|u - U\|_{W(0,t_m)}^2 \lesssim \mathcal{E}_0^2 + \sum_{n=1}^m \tau_n\big(\mathcal{E}_{\mathcal{G};n}^2(\mathcal{G}_n) + \mathcal{E}_{c\tau;n}^2\big)$$

$$+ \int_0^{t_m} \|f - f_n\|_*^2 \, dt$$

$$\lesssim \mathcal{E}_0^2 + \|u - U\|_{W(0,t_m)}^2 + \int_0^{t_m} \|f - f_n\|_*^2 \, dt$$

$$+ \sum_{n=1}^m \tau_n \operatorname{osc}(U_{n-1}, U_n, \mathcal{G}_n).$$

Proof. The first estimate is just the upper bound given in Theorem 3.4.6. The second estimate follows directly by summing the local lower bound given in Theorem 3.4.13 for $n = 1, \ldots, m$. $\square$

Finally, the upper and lower bounds in Corollary 3.4.15 justify the naming "error indicators" of the quantities introduced in (3.16). In particular, considering the estimate of the consistency error given in (3.24) together with Corollary 3.4.15 we see that the quantity

$$\mathcal{E}_0^2 + \sum_{n=1}^N \tau_n\big(\mathcal{E}_{\mathcal{G};n}^2(\mathcal{G}_n) + \mathcal{E}_{c\tau;n}^2 + \mathcal{E}_{f;n}^2\big)$$

constitutes a *reliable* and *efficient* error estimator since it bounds the true discretization error from above and below (up to oscillation) while being easily computable. We say that the error estimator is reliable as it bounds the error from above meaning that a small estimator guarantees a small error. Moreover, the error estimator is efficient since it bounds the error from below meaning that the estimator may only be big if the true error is big as well.

Chapter 4

The adaptive algorithm ASTFEM

In this chapter, we present an **A**daptive **S**pace **T**ime **F**inite **E**lement **M**ethod (ASTFEM) for solving the parabolic equation (2.1), which was also considered in [27]. ASTFEM is based on the Euler-Galerkin discretization (EG), which we recall reads: For given $U_0 \in \mathbb{V}_0$ find $U_n \in \mathbb{V}_n$, $n = 1, \ldots, N$ such that

$$\frac{1}{\tau_n} \langle U_n - U_{n-1}, V \rangle + \mathcal{B}(U_n, V) = \langle f_n, V \rangle \qquad \text{for all } V \in \mathbb{V}_n.$$
(EG)

The ultimate aim of such an algorithm is to compute a sequence of solutions $U_n \in \mathbb{V}_n$, $n = 1, \ldots, N$, such that the corresponding linear interpolation U satisfies the error bound

$$\|u - U\|_{\mathbb{W}(0,T)}^2 \leq \text{TOL}$$
(4.1)

for a given positive tolerance $\text{TOL} > 0$. Moreover, as we are interested in an *adaptive* algorithm, we demand that the algorithm autonomously generates the time step sizes τ_n as well as the finite element spaces $\mathbb{V}_n$. To be more precise, for the latter requirement we demand that the algorithm creates a mesh $\mathcal{G}_n$ for each time step n underlying the finite element space $\mathbb{V}_n = \mathbb{V}(\mathcal{G}_n)$. As we will see in the following, both the time step sizes and the meshes will be generated iteratively. Hence, we also have to address the

---

Algorithm 2 Basic time stepping

---

while $t_n < T$ **do**

 Increase $n := n + 1$;

 Generate time step size τ_n and finite element space $\mathbb{V}_n = \mathbb{V}(\mathcal{G}_n)$;

 Solve (EG) for $U_n \in \mathbb{V}_n$;

 Set new time node $t_n := t_{n-1} + \tau_n$;

end while

---

question of termination. Therefore, we have to consider the termination of those iterative processes yielding the termination of any *single* time step. On the other hand, we must keep in mind that an arbitrary sequence of time step sizes τ_n not necessarily implies that the desired final time T will be reached as premature convergence $\sum_{n=1}^{\infty} \tau_n = T' < T$ may occur. We shall focus on these concerns in Section 4.3. Summarizing, the demands on an abstract adaptive algorithm for solving (EG) are threefold:

(1) The treatment of any single time step shall terminate;

(2) The final time T shall be reached;

(3) The tolerance $\|u - U\|_{\mathbb{W}(0,T)}^2 \leq \texttt{TOL}$ shall be met.

 Since requirements (1) and (2) are closely linked to the actual algorithm, we first motivate its abstract structure. As suggested by the iterative structure of (EG), the algorithm shall solve equations (EG) step by step. Starting from a given initial approximation $U_0 \in \mathbb{V}_0$ of u_0, this requires the generation of a time step size τ_1 and a grid $\mathcal{G}_1$ giving rise to the finite element space $\mathbb{V}_1$ before actually solving (EG) for $n = 1$. Repeating this procedure for the subsequent time steps, we arrive at the basic abstract time stepping algorithm stated in Algorithm 2, which we start with $n := 0$ and $t_n := 0$. In the process, the generation of the new time step size and the new finite element space are subject to demand (3) above. That is, both time step sizes and finite element spaces shall be chosen such that the overall error meets its tolerance as stated in (4.1). Note that the generation of the next time step size and the next finite element space are pooled in Algorithm 2. This is necessary since modifying the time step size changes the (stationary) problem in (EG) which in turn may require a different finite element space. We will elaborate on this coupling in Section 4.2.6.

Structure of the chapter The basic structure of this chapter is to first consider general ways of controlling the discretization error (Section 4.1) which eventually leads to strategies for reducing errors from different error sources (Section 4.2). These strategies are the motivation for the actual principle of error control employed in ASTFEM (Section 4.3). Based on this principle, we introduce the abstract structure of ASTFEM (Section 4.4) and show that this abstract algorithm terminates within tolerance (Section 4.5). Only then, we present a concrete realization of ASTFEM which fits into the developed framework (Section 4.6).

For this important chapter we also provide a more detailed outline: In Section 4.1, we present two error control strategies A and B to actually establish an overall tolerance as demanded in (4.1) using the a posteriori error estimate from Theorem 3.4.15. Most importantly, error control strategy B allows for using a *global* minimal time step size $\tau_* > 0$, which turns out to be the key ingredient for showing that ASTFEM eventually terminates and especially reaches the desired final time T. To allow for a minimal time step size while guaranteeing that any demanded tolerance will be reached, error control strategy B makes essential use of a uniform global energy estimate introduced in Section 4.1.3, which is a generalization of a similar energy estimate originally derived in [27]. Both error control strategies A and B demand that certain criteria are satisfied in each individual time step.

In Section 4.2, we thus consider how the various criteria providing error control from Section 4.1 can be achieved. In light of these considerations, we point out the pros and cons of the two error control strategies and derive yet another error control strategy uniting their strong points in Section 4.3. This error control will be employed in ASTFEM as in [27].

In Section 4.4, the structure of ASTFEM is motivated and presented. To that end, we introduce abstract modules which we combine to give rise to ASTFEM. In Section 4.5 we then exploit properties of theses modules to show that ASTFEM terminates and reaches any given positive tolerance. The key ingredient here is the new error control strategy C which allows for using a global minimal time step size while obeying the demanded tolerance.

In Section 4.6 we present realizations of the abstract modules introduced earlier and show that these concrete implementations possess the properties demanded of the abstract modules.

4.1 Principles of error control

In this section, we will employ the error indicators presented and analyzed in Section 3.4 to establish the error bound demanded in (4.1). This will eventually require to adopt the time step sizes and the triangulations respectively the finite element spaces. These changes will be carried out iteratively in each single time step as suggested above. To keep the presentation precise and transparent, we equip the error indicators introduced in (3.16) with explicit arguments in the following subsection and hence make them more flexible.

4.1.1 Re-definition of error indicators

For a triangulation $\mathcal{G}$ and a function $v \in L^2(\Omega)$ we define the general version of the initial error indicator as

$$\mathcal{E}_0^2(v, \mathcal{G}) := 3 \left\| v - \mathcal{P}_\mathcal{G} v \right\|_{L^2(\Omega)}^2, \qquad (4.2a)$$

where $\mathcal{P}_\mathcal{G} : L^2(\Omega) \to \mathbb{V}(\mathcal{G})$ is an arbitrary "approximation operator" which produces an approximation of $v \in L^2(\Omega)$ in the finite element space $\mathbb{V}(\mathcal{G})$, see Remark 4.1.1. Furthermore, for $v, w \in H_0^1(\Omega)$, $f \in L^2(0, T; L^2(\Omega))$, a finite element function $V \in \mathbb{V}(\mathcal{G})$ and times $t, \tau, t + \tau \in (0, T)$, we define the following argument-bearing versions of the consistency, coarsen-time and space error indicators:

$$\mathcal{E}_f^2(f, t, \tau) := C_f \inf_{\bar{g} \in L^2(\Omega)} \frac{1}{\tau} \int_t^{t+\tau} \left\| f - \bar{g} \right\|_{L^2(\Omega)}^2 dt, \qquad (4.2b)$$

$$\mathcal{E}_{c\tau}^2(v, w) := 5 \left\| v - w \right\|_\Omega^2, \qquad (4.2c)$$

$$\mathcal{E}_{\mathcal{G}}^2(V, v, \tau, \bar{f}, \mathcal{G}, E) := 15 \, C_h \left(h_E^2 \left\| \bar{f} - \frac{V - v}{\tau} - \mathcal{L}V \right\|_{L^2(E)}^2 \right.$$

$$\left. + h_E \left\| [\![\mathbf{A}\nabla V]\!] \right\|_{L^2(\partial E \cap \Omega)}^2 \right). \quad (4.2\mathrm{d})$$

Whereas we easily see that (4.2c) and (4.2d) coincide with the coarsen-time respectively the space indicator from (3.16) provided the appropriate arguments, (4.2b) is of different structure than the consistency error indicator in (3.16). For that reason, we note that the infimum in (4.2b) is attained for $\bar{g} := \frac{1}{\tau} \int_t^{t+\tau} f \, dt$ and hence, the consistency error indicator coincides with its previous version as well. Finally, to enhance readability we abbreviate

$$\mathcal{E}_{\mathcal{G}}^2(V, v, \tau, \bar{f}, \mathcal{G}) := \sum_{E \in \mathcal{G}} \mathcal{E}_{\mathcal{G}}^2(V, v, \tau, \bar{f}, \mathcal{G}, E).$$

As suggested by the labeling as initial, consistency, coarsen-time and space indicator, the four indicators in (4.2) are related to different error sources. Such a splitting is very convenient for an adaptive algorithm since it allows for taking appropriate measures for reducing individual indicators, see Section 4.2.

Remark 4.1.1 (Approximation of u_0)
For a general $v \in L^2(\Omega)$ possessing no further regularity, a natural choice for the "approximation operator" is the L^2-projection, i. e., $\mathcal{P}_{\mathcal{G}} := \Pi_{\mathcal{G}}$. However, in many applications, additional regularity is provided. Particularly, for (piecewise) continuous functions, we may employ the Lagrange interpolation operator $\mathcal{P}_{\mathcal{G}} := \mathcal{I}_{\mathbb{V}(\mathcal{G})}$ of Theorem 3.1.16.

We now want to employ those error indicators together with the upper bound presented in Theorem 3.4.6 and Corollary 3.4.7 to satisfy the tolerance criterion stated in (4.1). With above re-definition of the error indicators, the upper bound from Corollary 3.4.7 reads

$$\|u - U\|_{\mathbb{W}(0,T)}^2 \leq \mathcal{E}_0^2(u_0, \mathcal{G}_0) + \sum_{n=1}^N \tau_n \left(\mathcal{E}_{\mathcal{G}}^2(U_n, U_{n-1}, \tau_n, f_n, \mathcal{G}_n) \right.$$

$$\left. + \mathcal{E}_{c\tau}^2(U_n, U_{n-1}) + \mathcal{E}_f^2(f, t_{n-1}, \tau_n) \right)$$

$$(4.3)$$

and we exploit this error bound to ultimately satisfy the overall tolerance (4.1).

4.1.2 Uniform bound on indicators

One possibility to bound $\|u - U\|^2_{\mathrm{W}(0,T)}$ involves bounding the spatial, coarsen-time and consistency error indicators uniformly for each time step n. To be more precise, we split the overall tolerance TOL into data tolerances $\mathrm{TOL}_0 > 0$ and $\mathrm{TOL}_f > 0$, and tolerances for the space and coarsen-time indicators $\mathrm{TOL}_\mathcal{G} > 0$ and $\mathrm{TOL}_{\mathrm{ct}} > 0$ respectively. For the data related indicators we then demand

$$\mathcal{E}_0^2(u_0, \mathcal{G}_0) \leq \mathrm{TOL}_0 \tag{4.4a}$$

and in every time step

$$\mathcal{E}_f^2(f, t_{n-1}, \tau_n) \leq \mathrm{TOL}_f. \tag{4.4b}$$

For the space and coarsen-time indicators, we demand in every time step

$$\mathcal{E}_\mathcal{G}^2(U_n, U_{n-1}, \tau_n, f_n, \mathcal{G}_n) \leq \mathrm{TOL}_\mathcal{G}, \tag{4.4c}$$

$$\mathcal{E}_{c\tau}^2(U_n, U_{n-1}) \leq \mathrm{TOL}_{\mathrm{ct}}. \tag{4.4d}$$

Once the four requests in (4.4) are satisfied, estimate (4.3) provides us with error control in the sense that

$$\|u - U\|^2_{\mathrm{W}(0,T)} \leq \mathrm{TOL}_0 + \sum_{n=1}^{N} \tau_n \big(\mathrm{TOL}_\mathcal{G} + \mathrm{TOL}_{\mathrm{ct}} + \mathrm{TOL}_f\big). \tag{4.5}$$

To reach the desired error bound (4.1), it only remains to choose the individual tolerances TOL_0, TOL_f, $\mathrm{TOL}_\mathcal{G}$ and $\mathrm{TOL}_{\mathrm{ct}}$ such that

$$\mathrm{TOL}_0 + \sum_{n=1}^{N} \tau_n \big(\mathrm{TOL}_\mathcal{G} + \mathrm{TOL}_{\mathrm{ct}} + \mathrm{TOL}_f\big) \leq \mathrm{TOL}, \tag{4.6}$$

which we always assume in what follows. Condition (4.6) on the individual tolerances may for example be accomplished by setting

$$\mathrm{TOL}_0 := \frac{\mathrm{TOL}}{4}, \quad \mathrm{TOL}_\mathcal{G} := \frac{\mathrm{TOL}}{4T}, \quad \mathrm{TOL}_{\mathrm{ct}} := \frac{\mathrm{TOL}}{4T}, \quad \mathrm{TOL}_f := \frac{\mathrm{TOL}}{4T}.$$

Concluding, we note that the overall error bound (4.1) is reached if the individual tolerances are chosen appropriately and the bounds in (4.4) are satisfied. We refer to this kind of error control as error control A:

Error control A
For a given tolerance TOL > 0, let the individual bounds stated in (4.4) be valid at the end of each time step $n \geq 0$ with subtolerances satisfying (4.6). Then for $t_N = T$ the bound

$$\|u - U\|^2_{\mathbb{W}(0,T)} \leq \textit{TOL}$$

is reached.

As we shall see later (cf. Section 4.2.5), it is appropriate to split $\mathcal{E}^2_{c\tau}(U_n, U_{n-1})$ via triangle inequality and Young's inequality into

$$\mathcal{E}^2_{c\tau}(U_n, U_{n-1}) = 5 \|U_n - U_{n-1}\|^2_\Omega$$
$$\leq 10 \|U_n - \Pi_n U_{n-1}\|^2_\Omega + 10 \|\Pi_n U_{n-1} - U_{n-1}\|^2_\Omega, \quad (4.7)$$

where $\Pi_n : H^1_0(\Omega) \to \mathbb{V}_n$ denotes the L^2-projection onto $\mathbb{V}_n$. Conveniently we define for $v, w \in H^1_0(\Omega)$, a triangulation $\mathcal{G}$ and an element $E \in \mathcal{G}$ the two indicators

$$\mathcal{E}^2_\tau(v, w, \mathcal{G}) := 10 \|v - \Pi_\mathcal{G} w\|^2_\Omega, \quad (4.8a)$$
$$\mathcal{E}^2_c(w, \mathcal{G}, E) := 10 \|\Pi_\mathcal{G} w - w\|^2_E, \quad (4.8b)$$

and as before we abbreviate $\mathcal{E}^2_c(w, \mathcal{G}) := \sum_{E \in \mathcal{G}} \mathcal{E}^2_c(w, \mathcal{G}, E)$. As motivated later in Section 4.2.5, it is suitable to name those indicators *time* and *coarsen* indicator respectively.

With splitting (4.7), we may substitute $\sum^N_{n=1} \tau_n \mathcal{E}^2_{c\tau}(U_n, U_{n-1})$ in error estimate (4.3) in terms of the newly defined time and coarsen indicators to attain

$$\|u - U\|^2_{\mathbb{W}(0,T)} \leq \mathcal{E}^2_0(u_0, \mathcal{G}_0) + \sum^N_{n=1} \tau_n \left(\mathcal{E}^2_\mathcal{G}(U_n, U_{n-1}, \tau_n, f_n, \mathcal{G}_n) \right.$$

$$\left. + \mathcal{E}^2_f(f, t_{n-1}, \tau_n) + \mathcal{E}^2_\tau(U_n, U_{n-1}, \mathcal{G}_n) + \mathcal{E}^2_c(U_{n-1}, \mathcal{G}_n) \right).$$

Mimicking the developments above, error control can then be provided by demanding for every time step $n = 1, \ldots, N$

$$\mathcal{E}_\tau^2(U_n, U_{n-1}, \mathcal{G}_n) \leq \text{TOL}_t, \qquad (4.9a)$$

$$\mathcal{E}_c^2(U_{n-1}, \mathcal{G}_n) \leq \text{TOL}_c, \qquad (4.9b)$$

with tolerances $\text{TOL}_t > 0$, $\text{TOL}_c \geq 0$ instead of (4.4d). Note that whereas we have to choose a positive tolerance TOL_t for the time error since we cannot guarantee that a zero time error indicator can be achieved, the tolerance TOL_c for the coarsening error can actually be chosen as zero, as we are indeed able to provide zero coarsening error, cf. Section 4.2.5.

In case of (4.9), error estimate (4.3) implies as above

$$\|u - U\|_{\mathbb{W}(0,T)}^2 \leq \text{TOL}_0 + \sum_{n=1}^{N} \tau_n \big(\text{TOL}_\mathcal{G} + \text{TOL}_t + \text{TOL}_c + \text{TOL}_f\big).$$

This way of error control is closely related to error control A and we hence refer to it as error control A':

Error control A'

For a given tolerance TOL > 0, let the individual bounds stated in (4.4a)–(4.4c) and (4.9) be valid at the end of each time step $n \geq 0$ with sub-tolerances satisfying $TOL_0 + \sum_{n=1}^{N} \tau_n \big(TOL_\mathcal{G} + TOL_t + TOL_c + TOL_f\big) \leq TOL$. Then for $t_N = T$ the bound

$$\|u - U\|_{\mathbb{W}(0,T)}^2 \leq TOL$$

is reached.

Note that error control A and A' only minorly differ from each other: Whereas error control A directly imposes a bound on the coarsen-time indicator $\mathcal{E}_{c\tau}^2(U_n, U_{n-1})$, error control A' uses control of the two sub-indicators $\mathcal{E}_\tau^2(U_n, U_{n-1}, \mathcal{G}_n)$ and $\mathcal{E}_c^2(U_{n-1}, \mathcal{G}_n)$. The philosophy behind both strategies however is the same: A sum of the type $\sum_{n=1}^{N} \tau_n \zeta_n$ is controlled by imposing a uniform bound on each single constituent, i. e., $\zeta_n \leq C$ for all $n = 1, \ldots, N$. In particular, this does *not* allow for any term $\zeta_{n_1} > C$ which could well be compensated by another term $\zeta_{n_2} \ll C$.

4.1.3 Error control via energy estimate

In this section, we particularly focus on the time discretization error $\sum_{n=1}^{N} \tau_n \mathcal{E}_\tau^2(U_n, U_{n-1}, \mathcal{G}_n)$ and introduce an alternate way of bounding it. Whereas above approach imposes a specific bound on $\mathcal{E}_\tau^2(U_n, U_{n-1}, \mathcal{G}_n)$ in each time step n, we now establish a uniform bound on $\sum_{n=1}^{N} \mathcal{E}_\tau^2(U_n, U_{n-1}, \mathcal{G}_n)$ emphasizing that *no* time step sizes are involved in this sum. A similar bound for symmetric problems was originally derived by Kreuzer, Möller, Siebert, and Schmidt in [27]. The term "uniform" has to be understood in the sense that the bound is independent of the sequence $\{\tau_n\}_{n=1,\ldots,N}$ of used time step sizes. This uniformity leaves us at liberty to employ the time step sizes τ_n as scaling factors to ensure a certain bound on $\sum_{n=1}^{N} \tau_n \mathcal{E}_\tau^2(U_n, U_{n-1}, \mathcal{G}_n)$. This principle of error control is essential for the termination of ASTFEM as it ultimately allows for using a globally minimal time step size.

Before presenting a uniform global energy estimate in Proposition 4.1.3, which is the key ingredient for the alternate error control introduced in this section, we need the following lemma.

Lemma 4.1.2
Let $\mathcal{C}_P$ denote the Poincaré-Friedrichs constant from Lemma 2.1.3 and let $U_0 \in \mathbb{V}_0$ be a given approximation of u_0. Further, let $U_n \in \mathbb{V}_n$, $n = 1, \ldots, N$, be the solutions to (EG) with corresponding time step sizes $\tau_1, \ldots, \tau_N$. Then for any $m = 1, \ldots, N$, the following estimate holds:

$$\|U_m\|^2 + \sum_{n=1}^{m} \left(\|U_n - U_{n-1}\|^2 + \tau_n \|U_n\|_\Omega^2 \right) \leq \|U_0\|^2 + \mathcal{C}_P^2 \sum_{n=1}^{m} \tau_n \|f_n\|^2 .$$

In particular, we have

$$\sum_{n=1}^{m} \tau_n \|U_n\|_\Omega^2 \leq \|U_0\|^2 + \mathcal{C}_P^2 \sum_{n=1}^{m} \tau_n \|f_n\|^2 .$$

Proof. We choose $V := U_n \in \mathbb{V}_n$ as a test function in (EG) to obtain

$$\frac{1}{\tau_n} \langle U_n - U_{n-1} , U_n \rangle + \|U_n\|_\Omega^2 = \langle f_n , U_n \rangle,$$

which we equivalently write as

$$\|U_n\|^2 - \langle U_{n-1} , U_n \rangle + \tau_n \|U_n\|_\Omega^2 = \tau_n \langle f_n , U_n \rangle. \tag{4.10}$$

From the general identity $\|v - w\|^2 = \|v\|^2 + \|w\|^2 - 2\langle v, w\rangle$ for all $v, w \in H_0^1(\Omega)$ we deduce

$$\langle U_{n-1}, U_n\rangle = \frac{1}{2}\left(\|U_n\|^2 + \|U_{n-1}\|^2 - \|U_n - U_{n-1}\|^2\right)$$

to obtain from equation (4.10)

$$\frac{1}{2}\|U_n\|^2 - \frac{1}{2}\|U_{n-1}\|^2 + \frac{1}{2}\|U_n - U_{n-1}\|^2 + \tau_n\|U_n\|_\Omega^2 = \tau_n\langle f_n, U_n\rangle.$$
$$(4.11)$$

Using Cauchy-Schwarz and Young's inequalities, we estimate for the right hand side for an arbitrary $\delta > 0$

$$\tau_n\langle f_n, U_n\rangle \leq \tau_n\|f_n\|\|U_n\| \leq \frac{\delta}{2}\tau_n\|f_n\|^2 + \frac{1}{2\delta}\tau_n\|U_n\|^2.$$

Employing the Poincaré-Friedrichs inequality from Lemma 2.1.3 and choosing $\delta := \mathcal{C}_\mathcal{P}^2$, we further find

$$\leq \frac{\mathcal{C}_\mathcal{P}^2}{2}\tau_n\|f_n\|^2 + \frac{\tau_n}{2}\|U_n\|_\Omega^2.$$

With this estimate, equation (4.11) implies

$$\|U_n\|^2 - \|U_{n-1}\|^2 + \|U_n - U_{n-1}\|^2 + \tau_n\|U_n\|_\Omega^2 \leq \tau_n\mathcal{C}_\mathcal{P}^2\|f_n\|^2$$

and summing up for $n = 1, \ldots, m$, we find

$$\|U_m\|^2 + \sum_{n=1}^{m}\left(\|U_n - U_{n-1}\|^2 + \tau_n\|U_n\|_\Omega^2\right) \leq \|U_0\|^2 + \mathcal{C}_\mathcal{P}^2\sum_{n=1}^{m}\tau_n\|f_n\|^2.$$

$$\square$$

In the following uniform global energy estimate, we explicitly consider the special case of *symmetric* bilinear forms $\mathcal{B}$ since in this case, we get a constant-free estimate, cf. [27]. The bound on $\sum_{n=1}^{m}\tau_n\|U_n\|_\Omega^2$ given in the foregoing lemma is not needed in this special case but only in the general case of non-symmetric $\mathcal{B}$. We also refer to the continuous analog stated in Theorem 2.3.3 as well as the associated Remark 2.3.4.

Proposition 4.1.3 (Uniform global energy estimate)
Let $U_0 \in \mathbb{V}_0$ be a given approximation of u_0 and let $U_n \in \mathbb{V}_n$, $n = 1, \ldots, N$, be the solutions to (EG) with corresponding time step sizes $\tau_1, \ldots, \tau_N$. Then for any $m = 1, \ldots, N$, the following estimate holds:

$$\sum_{n=1}^{m} \frac{1}{\tau_n} \|U_n - \Pi_n U_{n-1}\|^2 + \|U_n - \Pi_n U_{n-1}\|_{\Omega}^2 + \|U_n\|_{\Omega}^2 - \|\Pi_n U_{n-1}\|_{\Omega}^2$$

$$\leq 2\mathcal{C}_{\mathcal{B}_N}^2 \|U_0\|^2 + (2\mathcal{C}_{\mathcal{B}_N}^2 \mathcal{C}_{\mathcal{P}}^2 + 2) \sum_{n=1}^{m} \tau_n \|f_n\|^2 .$$

Beyond that, in case of a symmetric bilinear form $\mathcal{B}(\cdot, \cdot)$, we have the constant-free energy estimate

$$\sum_{n=1}^{m} \frac{1}{\tau_n} \|U_n - \Pi_n U_{n-1}\|^2 + \|U_n - \Pi_n U_{n-1}\|_{\Omega}^2 + \|U_n\|_{\Omega}^2 - \|\Pi_n U_{n-1}\|_{\Omega}^2$$

$$\leq \sum_{n=1}^{m} \tau_n \|f_n\|^2 .$$

Proof. First, we recall Lemma 2.3.1 to split the bilinear form $\mathcal{B}(\cdot, \cdot)$ into its symmetric and non-symmetric parts (where $\mathcal{B}_N(\cdot, \cdot) \equiv 0$ in case of symmetric $\mathcal{B}$)

$$\mathcal{B}(v, w) = \mathcal{B}_S(v, w) + \mathcal{B}_N(v, w) \qquad \text{for all } v, w \in H_0^1(\Omega).$$

With this splitting, (EG) reads equivalently

$$\frac{1}{\tau_n} \langle U_n - U_{n-1}, V \rangle + \mathcal{B}_S(U_n, V) = \langle f_n, V \rangle - \mathcal{B}_N(U_n, V) \quad (4.12)$$

for all $V \in \mathbb{V}_n$. We choose $V := U_n - \Pi_n U_{n-1} \in \mathbb{V}_n$ as a test function in (4.12). Producing the desired term $\|V\|_{\Omega}^2 = \mathcal{B}_S(V, V)$ on the left hand side, we get

$$\frac{1}{\tau_n} \|V\|^2 + \|V\|_{\Omega}^2 = \langle f_n, V \rangle - \mathcal{B}_S(\Pi_n U_{n-1}, V) - \mathcal{B}_N(U_n, V) \quad (4.13)$$

Using the same test function again in (4.12), we can also create the term $\|U_n\|_{\Omega}^2 = \mathcal{B}_S(U_n, U_n)$ on the left hand side:

$$\frac{1}{\tau_n} \|V\|^2 + \|U_n\|_{\Omega}^2 = \langle f_n, V \rangle + \mathcal{B}_S(U_n, \Pi_n U_{n-1}) - \mathcal{B}_N(U_n, V).$$

$$(4.14)$$

Adding up (4.13) and (4.14), we get with the symmetry and bilinearity of $\mathcal{B}_S(\cdot\,,\,\cdot)$

$$\frac{2}{\tau_n}\|V\|^2 + \|V\|_\Omega^2 + \|U_n\|_\Omega^2 = 2\langle f_n\,,\,V\rangle + \|\Pi_n U_{n-1}\|_\Omega^2 - 2\mathcal{B}_N(U_n\,,\,V).$$
(4.15)

Applying Cauchy-Schwarz and Young's inequalities, we estimate the first term on the right hand side via

$$2\langle f_n\,,\,V\rangle \le 2\tau_n\|f_n\|^2 + \frac{1}{2\tau_n}\|V\|^2.$$

Recalling Lemma 2.3.1, we also find for the last term of equation (4.15)

$$|2\mathcal{B}_N(U_n\,,\,V)| \le 2\mathcal{C}_{\mathcal{B}_N}\|U_n\|_\Omega\|V\| \le 2\mathcal{C}_{\mathcal{B}_N}^2\tau_n\|U_n\|_\Omega^2 + \frac{1}{2\tau_n}\|V\|^2.$$

Substituting these estimates into equation (4.15) and subtracting $\frac{1}{\tau_n}\|V\|^2$ on both sides, we derive

$$\frac{1}{\tau_n}\|V\|^2 + \|V\|_\Omega^2 + \|U_n\|_\Omega^2 \le 2\tau_n\|f_n\|^2 + 2\mathcal{C}_{\mathcal{B}_N}^2\tau_n\|U_n\|_\Omega^2$$
$$+ \|\Pi_n U_{n-1}\|_\Omega^2. \qquad (4.16)$$

Summing up (4.16) for $n = 1, \ldots, m$ yields

$$\sum_{n=1}^m \left(\frac{1}{\tau_n}\|U_n - \Pi_n U_{n-1}\|^2 + \|U_n - \Pi_n U_{n-1}\|_\Omega^2 \right.$$
$$+ \|U_n\|_\Omega^2 - \|\Pi_n U_{n-1}\|_\Omega^2 \Big)$$
$$\le \sum_{n=1}^m \left(2\tau_n\|f_n\|^2 + 2\mathcal{C}_{\mathcal{B}_N}^2\tau_n\|U_n\|_\Omega^2 \right)$$

and employing Lemma 4.1.2 we further estimate

$$\le 2\mathcal{C}_{\mathcal{B}_N}^2\|U_0\|^2 + (2\mathcal{C}_{\mathcal{B}_N}^2\mathcal{C}_{\mathcal{P}}^2 + 2)\sum_{n=1}^m \tau_n\|f_n\|^2.$$

Hereby, we established the claim for general bilinear forms $\mathcal{B}$. In case of symmetric $\mathcal{B}$ however, we have $\mathcal{B}_N(\cdot\,,\,\cdot) \equiv 0$ and thus, equation (4.15) implies using Cauchy-Schwarz and Young's inequality

$$\frac{2}{\tau_n} \|V\|^2 + \|V\|_\Omega^2 + \|U_n\|_\Omega^2 = 2\langle f_n\,,\,V\rangle + \|\Pi_n U_{n-1}\|_\Omega^2$$

$$\leq \tau_n \|f_n\|^2 + \frac{1}{\tau_n} \|V\|^2 + \|\Pi_n U_{n-1}\|_\Omega^2\,.$$

Adding $-\frac{1}{\tau_n} \|V\|^2$ on both sides and summing up for $n = 1, \ldots, m$ proves the additional claim for symmetric $\mathcal{B}$. $\qquad\square$

Remark 4.1.4 (Uniform global energy estimate for symmetric $\mathcal{B}$) We particularly point out that in case of a symmetric bilinear form $\mathcal{B}(\cdot\,,\,\cdot)$, the bound given in Proposition 4.1.3 is sharp in the sense that equality may be reached. This can be seen by observing that for proving the constant-free bound for symmetric bilinear forms only Cauchy-Schwarz inequality — which indeed is sharp — is used.

As indicated, Proposition 4.1.3 is the key ingredient for establishing a bound on the sum of time error indicators given by $\sum_{n=1}^{N} \mathcal{E}_\tau^2(U_n, U_{n-1}, \mathcal{G}_n)$ which is uniform in the sense that it does not depend on the employed sequence of time step sizes.

Corollary 4.1.5 (Uniform bound for time error indicators) *Let $U_0 \in \mathbb{V}_0$ be a given approximation of u_0 and for $n = 1, \ldots, N$, let $U_n \in \mathbb{V}_n$ be the sequence of solutions to (EG) with corresponding time step sizes τ_n. Assume that for $n = 1, \ldots, N$*

$$\|U_{n-1}\|_\Omega^2 - \|\Pi_n U_{n-1}\|_\Omega^2 + \frac{1}{\tau_n} \|U_n - \Pi_n U_{n-1}\|^2 \geq 0. \qquad (4.17)$$

Then it holds the estimate

$$\sum_{n=1}^{m} \|U_n - \Pi_n U_{n-1}\|_\Omega^2 \leq \|U_0\|_\Omega^2 + 2\mathcal{C}_{\mathcal{B}_N}^2 \|U_0\|^2$$

$$+ (2\mathcal{C}_{\mathcal{B}_N}^2 \mathcal{C}_{\mathcal{P}}^2 + 2) \|f\|_{L^2(0,t_m;L^2(\Omega))}^2 - \|U_m\|_\Omega^2\,.$$

Beyond that, in case of a symmetric bilinear form $\mathcal{B}(\cdot\,,\,\cdot)$, we have the following constant-free estimate:

$$\sum_{n=1}^{m} \|U_n - \Pi_n U_{n-1}\|_\Omega^2 \leq \|U_0\|_\Omega^2 + \|f\|_{L^2(0,t_m;L^2(\Omega))}^2 - \|U_m\|_\Omega^2\,.$$

In any case, particularly the sum of the time error indicators $\sum_{n=1}^{N} \mathcal{E}_{\mathcal{T}}^2(U_n, U_{n-1}, \mathcal{G}_n) = \sum_{n=1}^{N} 10 \, \|U_n - \Pi_n U_{n-1}\|_{\Omega}^2$ *is uniformly bounded irrespective of the sequence of time step sizes used.*

Proof. Summing up the nonnegative terms in (4.17) yields

$$
0 \le \sum_{n=1}^{m} \|U_{n-1}\|_{\Omega}^2 - \|\Pi_n U_{n-1}\|_{\Omega}^2 + \frac{1}{\tau_n} \|U_n - \Pi_n U_{n-1}\|^2 ,
$$

which is equivalent to

$$
\|U_m\|_{\Omega}^2 - \|U_0\|_{\Omega}^2 \le \sum_{n=1}^{m} \|U_n\|_{\Omega}^2 - \|\Pi_n U_{n-1}\|_{\Omega}^2 + \frac{1}{\tau_n} \|U_n - \Pi_n U_{n-1}\|^2 .
$$

Hence, we have

$$
\sum_{n=1}^{m} \|U_n - \Pi_n U_{n-1}\|_{\Omega}^2 + \|U_m\|_{\Omega}^2 - \|U_0\|_{\Omega}^2
$$

$$
\le \sum_{n=1}^{m} \|U_n - \Pi_n U_{n-1}\|_{\Omega}^2 + \|U_n\|_{\Omega}^2 - \|\Pi_n U_{n-1}\|_{\Omega}^2
$$

$$
+ \frac{1}{\tau_n} \|U_n - \Pi_n U_{n-1}\|^2 . \tag{4.18}
$$

Employing the estimate for general bilinear forms of Proposition 4.1.3, this directly yields

$$
\sum_{n=1}^{m} \|U_n - \Pi_n U_{n-1}\|_{\Omega}^2 \le \|U_0\|_{\Omega}^2 - \|U_m\|_{\Omega}^2 + 2\mathcal{C}_{\mathcal{B}_N}^2 \|U_0\|^2
$$

$$
+ (2\mathcal{C}_{\mathcal{B}_N}^2 \mathcal{C}_{\mathcal{P}}^2 + 2) \sum_{n=1}^{m} \tau_n \|f_n\|^2 ,
$$

and observing that $\sum_{n=1}^{m} \tau_n \|f_n\|^2 \le \|f\|_{L^2(0,t_m;L^2(\Omega))}^2$ we find

$$
\sum_{n=1}^{m} \|U_n - \Pi_n U_{n-1}\|_{\Omega}^2 \le \|U_0\|_{\Omega}^2 - \|U_m\|_{\Omega}^2
$$

$$
+ 2\mathcal{C}_{\mathcal{B}_N}^2 \|U_0\|^2 + (2\mathcal{C}_{\mathcal{B}_N}^2 \mathcal{C}_{\mathcal{P}}^2 + 2) \|f\|_{L^2(0,t_m;L^2(\Omega))}^2 .
$$

Analogously, using the estimate for symmetric bilinear forms of Proposition 4.1.3 in (4.18) we find

$$\sum_{n=1}^{m} \|U_n - \Pi_n U_{n-1}\|_{\Omega}^2 \leq \|U_0\|_{\Omega}^2 - \|U_m\|_{\Omega}^2 + \|f\|_{L^2(0,t_m;L^2(\Omega))}^2 .$$

$\square$

Remark 4.1.6 (Nested spaces)
In case of nested finite element spaces $\mathbb{V}_{n-1} \subset \mathbb{V}_n$ for all $n = 1, \ldots, N$, condition (4.17) in above corollary is apparently always satisfied, since in this case, $\Pi_n : \mathbb{V}_{n-1} \subset \mathbb{V}_n \to \mathbb{V}_n$ is the identity. Corollary 4.1.5 then unconditionally bounds the sum of all energy jumps $\|U_n - U_{n-1}\|_{\Omega}^2$ between consecutive solutions by exclusively data dependent terms.

Remark 4.1.7 (Energy gain)
In case of non-nested spaces $\mathbb{V}_{n-1} \not\subset \mathbb{V}_n$, the projection $\Pi_n U_{n-1}$ of U_{n-1} to the current finite element space $\mathbb{V}_n$ generally differs from $U_{n-1} \in \mathbb{V}_{n-1} \not\subset \mathbb{V}_n$. Since the L^2-projection $\Pi_n : \mathbb{V}_{n-1} \to \mathbb{V}_n$ does not match the energy norm $\|\cdot\|_{\Omega}$ — in the sense that it does *not* provide the estimate $\|\Pi_n U_{n-1}\|_{\Omega}^2 \leq \|U_{n-1}\|_{\Omega}^2$ — additional energy may be introduced by this projection, i.e., $\|\Pi_n U_{n-1}\|_{\Omega}^2 > \|U_{n-1}\|_{\Omega}^2$. In this light, condition (4.17) requires that this energy gain is not too big, particularly $\|\Pi_n U_{n-1}\|_{\Omega}^2 - \|U_{n-1}\|_{\Omega}^2 \leq \frac{1}{\tau_n} \|U_n - \Pi_n U_{n-1}\|^2$.

Since we will eventually employ Corollary 4.1.5 for error control, we conveniently define

$$\mathcal{E}_*^2(v, w, \tau, \mathcal{G}, E) := \frac{1}{\tau} \|v - \Pi_{\mathcal{G}} w\|_{L^2(E)}^2 + \|w\|_E^2 - \|\Pi_{\mathcal{G}} w\|_E^2 \quad (4.19)$$

and again, we abbreviate $\mathcal{E}_*^2(v, w, \tau, \mathcal{G}) := \sum_{E \in \mathcal{G}} \mathcal{E}_*^2(v, w, \tau, \mathcal{G}, E)$. With this notion, we conveniently rewrite condition (4.17) as

$$\mathcal{E}_*^2(U_n, U_{n-1}, \tau_n, \mathcal{G}_n) \geq 0 \quad (4.20)$$

for all $n = 1, \ldots, N$. Supposing condition (4.20) holds, Corollary 4.1.5 provides an a priori upper bound for the temporal error.

In particular, it holds for general bilinear forms

$$\sum_{n=1}^{N} \mathcal{E}_{\tau}^2(U_n, U_{n-1}, \mathcal{G}_n) \leq 10\Big(\|U_0\|_{\Omega}^2 + 2\mathcal{C}_{\mathcal{B}_N}^2 \|U_0\|^2$$

$$+ (2\mathcal{C}_{\mathcal{B}_N}^2 \mathcal{C}_{\mathcal{P}}^2 + 2) \|f\|_{L^2(0,T;L^2(\Omega))}^2 \Big). \qquad (4.21a)$$

In case of a symmetric bilinear form, beyond that we have the stronger constant-free estimate

$$\sum_{n=1}^{N} \mathcal{E}_{\tau}^2(U_n, U_{n-1}, \mathcal{G}_n) \leq 10\left(\|U_0\|_{\Omega}^2 + \|f\|_{L^2(0,T;L^2(\Omega))}^2 \right). \qquad (4.21b)$$

Since we are interested in controlling the sum of the time error indicators given by $\sum_{n=1}^{N} \tau_n \mathcal{E}_{\tau}^2(U_n, U_{n-1}, \mathcal{G}_n)$, we may hence use the time step sizes τ_n for scaling purposes as suggested above. In doing so, we directly obtain the following alternate error control by multiplying (4.21a) respectively (4.21b) with a fixed time step size τ_*.

Error control B
For an arbitrary tolerance $TOL_* > 0$ we have

$$\sum_{n=1}^{N} \tau_n \mathcal{E}_{\tau}^2(U_n, U_{n-1}, \mathcal{G}_n) \leq TOL_*$$

provided condition (4.20) is satisfied for all $n = 1, \ldots, N$, and all time step sizes are sufficiently small. This is we have

$$\tau_n \leq \tau_* \qquad \text{for all } n = 1, \ldots, N,$$

where for general bilinear forms, τ_* is defined as

$$\tau_* := \Big(10\Big(\|U_0\|_{\Omega}^2 + 2\mathcal{C}_{\mathcal{B}_N}^2 \|U_0\|^2$$

$$+ (2\mathcal{C}_{\mathcal{B}_N}^2 \mathcal{C}_{\mathcal{P}}^2 + 2) \|f\|_{L^2(0,T;L^2(\Omega))}^2 \Big)\Big)^{-1} TOL_* \qquad (4.22a)$$

with Poincaré constant $\mathcal{C}_{\mathcal{P}}$ from Lemma 2.1.3 and constant $\mathcal{C}_{\mathcal{B}_N}$ associated to $\mathcal{B}_N$ from Lemma 2.3.1. For symmetric bilinear forms, τ_* is defined as

$$\tau_* := \Big(10\Big(\|U_0\|_{\Omega}^2 + \|f\|_{L^2(0,T;L^2(\Omega))}^2 \Big)\Big)^{-1} TOL_*. \qquad (4.22b)$$

Error control B controls the term $\sum_{n=1}^{N} \tau_n \left\| U_n - \Pi_n U_{n-1} \right\|_{\Omega}^2$ by merely imposing conditions on the time step sizes τ_n in (4.22) and the energy gain $\mathcal{E}_*^2(U_n, U_{n-1}, \tau_n, \mathcal{G}_n)$ in (4.20). In particular, taking the energy gain criterion for granted, error control B introduces the *global minimal time step size* τ_* in the sense that choosing $\tau_n := \tau_*$ for all $n = 1, \ldots, N$ suffices for controlling the temporal error $\sum_{n=1}^{N} \tau_n \mathcal{E}_\tau^2(U_n, U_{n-1}, \mathcal{G}_n) \leq \text{TOL}_*$. We point out that the minimal time step size derived for the special case of *symmetric* $\mathcal{B}$ is indeed *larger* than the minimal time step size for general, potentially non-symmetric $\mathcal{B}$. Hence, explicitly treating the special case of symmetric $\mathcal{B}$ pays off.

Before commenting on the differences between error control A and B, we present an alternate point of view for Corollary 4.1.5.

Remark 4.1.8 (Integrability)
We consider the sequence $\{\left\| U_n - \Pi_n U_{n-1} \right\|_{\Omega}\}_{n=1,\ldots,N}$ as a piecewise constant function in time $k : [0, T] \to \mathbb{R}$ with

$$k(t) = \left\| U_n - \Pi_n U_{n-1} \right\|_{\Omega} \quad \text{for } t \in (t_{n-1}, t_n].$$

Note that for any sequence $\{\tau_n\}_{n \in \mathbb{N}}$ the function k is different in general but as we will see in the following, it holds $k \in L^2(0, T)$ independent of the sequence of time step sizes used. Further, we denote the sequence of time step sizes by $l \in \mathbb{R}_+^{\mathbb{N}}$. For an admissible sequence of time step sizes l we have $\|l\|_{\infty} \leq T$. By a superscript l we indicate the dependency on the sequence of time step sizes l. Since k^l is a piecewise constant function, we have

$$\sum_{n=1}^{N^l} \tau_n^l \left\| U_n^l - \Pi_n U_{n-1}^l \right\|_{\Omega}^2 = \int_0^T (k^l)^2(t) \, dt.$$

Provided the energy gain condition (4.20) is satisfied — which e. g., is always the case for nested finite element spaces (see Remark 4.1.7) — we obtain from Corollary 4.1.5 the estimate

$$\sum_{n=1}^{N^l} \left\| U_n^l - \Pi_n U_{n-1}^l \right\|_{\Omega}^2 \leq C$$

for all admissible $l \in \mathbb{R}_+^{\mathbb{N}}$. This leads to

$$\int_0^T (k^l)^2(t)\,dt = \sum_{n=1}^{N^l} \tau_n^l \left\| \left\| U_n^l - \Pi_n U_{n-1}^l \right\| \right\|_\Omega^2$$

$$\leq T \sum_{n=1}^{N^l} \left\| \left\| U_n^l - \Pi_n U_{n-1}^l \right\| \right\|_\Omega^2 \leq C.$$

Thus $k^l \in L^2(0,T)$ for all admissible time step sequences $l \in \mathbb{R}_+^{\mathbb{N}}$.

Remark 4.1.9 (Ways of temporal error control)
With regard to the temporal error, both error control A respectively A' and error control B aim at establishing a bound of the form

$$\sum_{n=1}^{N} \tau_n \mathcal{E}_\tau^2(U_n, U_{n-1}, \mathcal{G}_n) \leq \texttt{TOL} \tag{4.23}$$

for some tolerance $\texttt{TOL} > 0$. The approach taken in error control A respectively A' imposed a uniform bound on the time indicator $\mathcal{E}_\tau^2(U_n, U_{n-1}, \mathcal{G}_n)$ by demanding in each time step n

$$\mathcal{E}_\tau^2(U_n, U_{n-1}, \mathcal{G}_n) \leq \texttt{TOL}_{\text{t}}.$$

With this uniform bound installed we have

$$\sum_{n=1}^{N} \tau_n \mathcal{E}_\tau^2(U_n, U_{n-1}, \mathcal{G}_n) \leq \texttt{TOL}_{\text{t}} \sum_{n=1}^{N} \tau_n$$

and the convergence $\sum_{n=1}^{N} \tau_n = T$ yields the desired estimate of the form (4.23). Otherwise put, we estimated the L^2-type term $\sum_{n=1}^{N} \tau_n \mathcal{E}_\tau^2(U_n, U_{n-1}, \mathcal{G}_n)$ by imposing an L^∞-bound on the term $\mathcal{E}_\tau^2(U_n, U_{n-1}, \mathcal{G}_n)$, (cf. Remark 4.1.8).

Opposed to that, the approach taken in error control B establishes a uniform bound of the form

$$\sum_{n=1}^{N} \mathcal{E}_\tau^2(U_n, U_{n-1}, \mathcal{G}_n) \leq C.$$

As we may exploit this convergence of the sum of the time error indicators (opposed to the convergence of the sum of time step sizes in the approach of error control A respectively A'), we no longer rely on the convergence of the time step sizes and may thus employ τ_n as a scaling factor to ensure a specific bound.

4.2 Establishing error control

The discussion of principle ways of error control in the preceeding Section 4.1 resulted in the statement of certain criteria (given in (4.4) and (4.9) respectively (4.20) and (4.22)) which can be used to guarantee that an overall tolerance TOL > 0 is reached. However, we still have to deal with the question of how to satisfy those criteria. We will address this in the following by considering the individual error sources one at a time and stating abstract means for reducing them. These considerations are the motivation for the error control in ASTFEM which will be introduced in Section 4.3. In particular, here we do not provide any details on or properties of ASTFEM like e. g., termination. Rather, the modules of ASTFEM will only be presented later in Sections 4.4 and 4.6 and we will focus on termination and convergence of ASTFEM in Section 4.5.

To begin, we emphasize that the error bounds (4.4) and (4.9) used in error control A are a posteriori bounds. Consequently, the step

> "Generate time step size τ_n and finite element
> space $\mathbb{V}_n = \mathbb{V}(\mathcal{G}_n)$;"

in Algorithm 2 will contain iterative loops for determining the next time step size and finite element space. Likewise, we also aim at establishing the energy gain condition of (4.20) in an iterative process.

In order to clearly distinguish between solutions from one time step which are related to different time step sizes and/or finite element spaces, we introduce the iteration index k as further subscript to time step size, solution, mesh and finite element space. Particularly, the unique solution $U_{n,k} \in \mathbb{V}_{n,k}$ is related to precisely one time step size $\tau_{n,k}$ and one finite element space $\mathbb{V}_{n,k} := \mathbb{V}(\mathcal{G}_{n,k})$ as it is the solution of the corresponding version of (EG) given by

$$\frac{1}{\tau_{n,k}} \langle U_{n,k} - U_{n-1}, V \rangle + \mathcal{B}(U_{n,k}, V) = \langle f_{n,k}, V \rangle \quad \text{for all } V \in \mathbb{V}_{n,k}.$$

$$(\text{EG}_k)$$

Here, in analogy to the definition of f_n, we understand

$$f_{n,k} := \frac{1}{\tau_{n,k}} \int_{t_{n-1}}^{t_{n-1}+\tau_{n,k}} f(s) \, ds.$$

Using this notation, we analyze how the error control criteria from Section 4.1 can be reached. In the process, we first consider each criterion individually in Sections 4.2.1 through 4.2.5 before investigating coupling effects — which may arise while aiming for multiple criteria simultaneously — in Sections 4.2.6 and 4.2.7. Since we employed a space-time discretization for deriving (EG), the two available options for eventually satisfying the error control criteria from Section 4.1 are

(1) refining the mesh   and   (2) reducing the time step size.

We first examine the energy gain criterion which is important in the context of error control using the energy bound.

4.2.1   Energy gain

The energy gain criterion (4.20) demands

$$\mathcal{E}_*^2(U_n, U_{n-1}, \tau_n, \mathcal{G}_n) = \frac{1}{\tau_n} \left\| U_n - \Pi_n U_{n-1} \right\|^2$$
$$+ \left\| U_{n-1} \right\|_\Omega^2 - \left\| \Pi_n U_{n-1} \right\|_\Omega^2 \geq 0.$$

From Remark 4.1.6 we also recall that for nested finite element spaces $\mathbb{V}_{n-1} \subset \mathbb{V}_n$ we have $\Pi_n U_{n-1} = U_{n-1} \in \mathbb{V}_{n-1} \subset \mathbb{V}_n$ and thus

$$\mathcal{E}_*^2(U_n, U_{n-1}, \tau_n, \mathcal{G}_n) = \frac{1}{\tau_n} \left\| U_n - \Pi_n U_{n-1} \right\|^2 \geq 0.$$

Hence, we expect that refining the mesh $\mathcal{G}_n$ "towards" $\mathcal{G}_{n-1}$ is the appropriate means for eventually installing $\mathcal{E}_*^2(U_n, U_{n-1}, \tau_n, \mathcal{G}_n) \geq 0$. By "refining $\mathcal{G}_n$ towards $\mathcal{G}_{n-1}$" we understand generating a sequence of meshes $\{\mathcal{G}_{n,k}\}_{k \geq 0}$ starting with $\mathcal{G}_{n,0} = \mathcal{G}_n$ and producing refinements $\mathcal{G}_{n,k+1} \geq \mathcal{G}_{n,k}$ for $k \geq 0$, such that for some $K \geq 0$ we have $\mathcal{G}_{n,K} \geq \mathcal{G}_{n-1}$. For this $K \geq 0$ it holds $\mathbb{V}_{n-1} \subset \mathbb{V}(\mathcal{G}_{n,K})$ and thus the energy gain criterion is satisfied, i.e., more precisely we have $\mathcal{E}_*^2(U_{n,K}, U_{n-1}, \tau_n, \mathcal{G}_{n,K}) \geq 0$.

We mention that for this directed refinement we may *not* directly employ a standard mark-refine-procedure which only considers the set $\{\mathcal{E}_*^2(U_n, U_{n-1}, \tau_n, \mathcal{G}_n, E)\}_{E \in \mathcal{G}}$ of local "energy gain indicators" in order to select a set of marked elements. To understand this, we observe that the employed L^2-projection is a *global* operator, which particularly implies that the local energy gain indicator

$\mathcal{E}_*^2(U_n, U_{n-1}, \tau_n, \mathcal{G}_n, E)$ may be negative on an element E which was *not* coarsened from $\mathcal{G}_{n-1}$ to $\mathcal{G}_n$, i.e., $E \in \mathcal{G}_n$ and $E \in \mathcal{G}_{n-1}$. Thus, it is possible that the marking strategy only selects such un-coarsened elements and hence, it may happen that the actually coarsened elements are *not* refined. With these considerations, we realize that in order to get a directed refinement, we explicitly have to see for it that previously coarsened elements are refined again, cf. the paragraph on energy gain in Section 4.4.2.

We will now investigate how the error indicators involved in conditions (4.4) and (4.9) can be reduced to meet their respective bounds. Considering each indicator individually, we start with the indicators related to data resolution.

4.2.2 Initial error indicator

The condition on the initial error indicator from (4.4a) demands

$$\mathcal{E}_0^2(u_0, \mathcal{G}_0) \leq \mathrm{TOL}_0$$

and we recall from its definition that $\mathcal{E}_0^2(u_0, \mathcal{G}_0) = \|u_0 - \mathcal{P}_{\mathcal{G}_0} u_0\|^2$. Since we only assume L^2-regularity for the initial datum u_0, we do not have an interpolation estimate connecting the discretization error $\|u_0 - \mathcal{P}_{\mathcal{G}_0} u_0\|$ to the mesh $\mathcal{G}_0$. For this reason, we *assume* we have at hand a suitable approximation $U_0 := \mathcal{P}_{\mathcal{G}_0} u_0 \in \mathbb{V}(\mathcal{G}_0)$ of u_0 for some approximation operator $\mathcal{P}_{\mathcal{G}_0}$, cf. Remark 4.1.1. This is legitimate, since the process of generating U_0 is plain resolution of data.

Remark 4.2.1 (Approximation of initial data and convergence) In practical realizations, condition (4.4a) on the initial error indicator is achieved via adaptive approximation. That is, in the standard adaptive method (SEMR) (cf. Section 3.3.5), SOLVE does not actually solve any PDE but rather computes the approximation $U_0 = \mathcal{P}_{\mathcal{G}_0} u_0 \in \mathbb{V}(\mathcal{G}_0)$ of u_0 in the finite element space related to the current mesh. Moreover, the module ESTIMATE computes the actual approximation error $\|u_0 - \mathcal{P}_{\mathcal{G}_0} u_0\|$ as well as the local errors $\{\|u_0 - \mathcal{P}_{\mathcal{G}_0} u_0\|_{L^2(E)}\}_{E \in \mathcal{G}_0}$ on the current grid. Based on this information, a standard marking strategy together with the module REFINE adaptively refines the mesh. This modified version of (SEMR) is iterated until the demanded tolerance TOL_0 is

reached, cf. Algorithm 1 on page 77.
When employing the L^2-projection $\Pi_{\mathcal{G}_0} =: \mathcal{P}_{\mathcal{G}_0}$ to approximate u_0, convergence of this adaptive approximation can be obtained along the same lines as convergence for elliptic problems in [36] by employing the modified SOLVE and ESTIMATE modules, cf. Section 3.3.5. In the (common) special case of continuous initial data, we may wish to rather employ the Lagrange interpolation operator to approximate u_0 instead of above L^2-projection. For interpolation operators — which are local and thus quite different from the global projection operators — the basic convergence result of [36] cannot be applied directly since in particular, the Lagrange interpolation is not stable with respect to the L^2-norm. However, in most practical applications calling for approximating u_0 via an interpolation, additional regularity beyond $u_0 \in L^2(\Omega) \cap C(\overline{\Omega})$ is provided and guarantees that the adaptive approximation eventually reaches the desired tolerance.

4.2.3   Consistency error indicator

The condition on the consistency error indicator from (4.4b) demands

$$\mathcal{E}_f^2(f, t_{n-1}, \tau_n) \leq \mathrm{TOL}_f$$

and we recall from its definition that

$$\mathcal{E}_f^2(f, t_{n-1}, \tau_n) := \mathcal{C}_f \frac{1}{\tau_n} \int_{t_{n-1}}^{t_{n-1}+\tau_n} \|f(s) - f_n\|^2 \, ds$$

where $f_n = \frac{1}{\tau_n} \int_{t_{n-1}}^{t_{n-1}+\tau_n} f(s) \, ds$ is the mean value of f on the current time interval. We may understand this definition in the sense that $\mathcal{E}_f^2(f, t_{n-1}, \tau_n)$ represents the oscillation of f around its temporal mean f_n on the interval $(t_{n-1}, t_{n-1} + \tau_n)$. Thus, we expect that this indicator may be reduced by decreasing the time step size τ_n, which is confirmed by the following lemma.

Lemma 4.2.2 (Reduction of consistency error)
Assuming that f possesses the additional regularity $f \in H^1(0, T; L^2(\Omega))$, the consistency error indicator can be estimated by

$$\mathcal{E}_f^2(f, t, \tau) \leq \mathcal{C}_f \, \tau \, \|f'\|_{L^2(0,T;L^2(\Omega))}^2 \, .$$

In particular, the consistency indicator decreases (linearly) with the time step size τ.

Proof. We recall the definition of the consistency error indicator from (4.2b)

$$\mathcal{E}_f^2(f, t, \tau) = \frac{\mathcal{C}_f}{\tau} \int_t^{t+\tau} \left\| f(s) - \bar{f} \right\|^2 ds$$

with $\bar{f} = \frac{1}{\tau} \int_t^{t+\tau} f \, ds$. Making use of the additional regularity of f and employing a Poincaré argument, we thereby derive

$$\mathcal{E}_f^2(f, t, \tau) \leq \frac{\mathcal{C}_f}{\tau} \int_t^{t+\tau} \tau^2 \left\| f'(s) \right\|^2 ds \leq \mathcal{C}_f \tau \left\| f' \right\|_{L^2(0,T;L^2(\Omega))}^2.$$

$\square$

Assuming we have at hand an initial guess $\tau_{n,0} > 0$ for the time step size of the n-th time step, we state a basic algorithm which may be employed to establish (4.4b) in Algorithm 3. However, we anticipate that ASTFEM does *not* use the very basic Algorithm 3 but rather employs a more advanced module for controlling the consistency error (see Section 4.6.1 and particularly Algorithm 6 on page 160).

Algorithm 3 Basic consistency adaptation

Set $k := 0$;
while $\mathcal{E}_f^2(f, t_{n-1}, \tau_{n,k}) > \mathrm{TOL}_f$ **do**
    Set $k := k + 1$;
    Produce $\tau_{n,k}$ by reducing $\tau_{n,k-1}$;
end while
Return $\tau_n := \tau_{n,k}$;

Remark 4.2.3 (Position in overall algorithm)
We note that the consistency error indicator does not depend on the solution U_n in the n-th time step. This motivates to ensure bound (4.4b) on the consistency error before actually solving (EG). Moreover, we see from Lemma 4.2.2 that the consistency error does not increase as long as the time step size is not increased. This motivates to only allow for *reductions* of the time step size in the

remaining procedure of the current time step. In this case, once the bound (4.4b) is established, it remains valid throughout the current time step.

4.2.4 Space error indicator

The condition on the space error indicator from (4.4c) demands

$$\mathcal{E}_{\mathcal{G}}^2(U_n, U_{n-1}, \tau_n, f_n, \mathcal{G}_n) \leq \text{TOL}_{\mathcal{G}}$$

and we recall that $\mathcal{E}_{\mathcal{G}}^2(U_n, U_{n-1}, \tau_n, f_n, \mathcal{G}_n)$ is the error indicator of the associated stationary problem which reads: Find $u_n \in H_0^1(\Omega)$ such that

$$\mathcal{B}(u_n, v) + \frac{1}{\tau_n}\langle u_n, v\rangle = \langle f_n + \frac{1}{\tau_n}U_{n-1}, v\rangle \qquad \text{for all } v \in H_0^1(\Omega).$$

From Section 3.3 on elliptic problems, we know that the spatial indicator $\mathcal{E}_{\mathcal{G}}^2(U_n, U_{n-1}, \tau_n, \mathcal{G}_n)$ can be reduced by refining the triangulation underlying the finite element space $\mathbb{V}_n$ in the current time step. Thus, to satisfy criterion (4.4c), we may employ the standard solve-estimate-mark-refine iteration, which is implemented in Algorithm 1 on page 77. Notice however, that the underlying elliptic problem depends on the current time step size $\tau_{n,k}$. In particular, even if $\mathcal{E}_{\mathcal{G}}^2(U_{n,k}, U_{n-1}, \tau_{n,k}, \mathcal{G}_{n,k}) \leq \text{TOL}_{\mathcal{G}}$ we may well have $\mathcal{E}_{\mathcal{G}}^2(U_{n,k+1}, U_{n-1}, \tau_{n,k+1}, \mathcal{G}_{n,k+1}) > \text{TOL}_{\mathcal{G}}$ due to a change of the time step size $\tau_{n,k+1} \neq \tau_{n,k}$, see Section 4.2.6.

4.2.5 Coarsen-time error indicator

It remains to analyze the coarsen-time indicator, which is defined in (4.2c) as $\mathcal{E}_{c\tau}^2(U_n, U_{n-1}) = 5\|U_n - U_{n-1}\|_\Omega^2$ and shall meet the tolerance criterion (4.4d) which we recall reads

$$\mathcal{E}_{c\tau}^2(U_n, U_{n-1}) \leq \text{TOL}_{\text{ct}}.$$

For this indicator, refining the mesh may not be appropriate. The reason for that is that while U_{n-1} approximates the exact solution u at time t_{n-1}, U_n approximates the exact solution at a different time t_n. Since u is not constant in general, we cannot expect $\mathcal{E}_{c\tau}^2(U_n, U_{n-1})$ tending to zero as we refine the mesh.

On the other hand, simply reducing the time step size τ_n is also not sufficient in general. This is due to the fact that the finite element spaces $\mathbb{V}_{n-1} \not\subset \mathbb{V}_n$ are commonly not nested: With τ_n tending to zero, we would expect U_n to converge to U_{n-1}, but as $U_{n-1} \in \mathbb{V}_{n-1} \not\subset \mathbb{V}_n \ni U_n$, such a convergence is impossible. Thus we see that $\mathcal{E}_{c\tau}^2(U_n, U_{n-1})$ actually contains two error sources, one due to the gap between the non-nested spaces $\mathbb{V}_n$ and $\mathbb{V}_{n-1}$ (coarsening error) and one due to the time discretization. This justifies labeling the term coarsen-time indicator.

This observation motivates to split the coarsen-time indicator into a part related to coarsening and a part related to time discretization. We already used this splitting in Section 4.1, however, we did not provide a motivation there. We observe that for the L^2-projection $\Pi_{\mathcal{G}} : H_0^1(\Omega) \to \mathbb{V}(\mathcal{G})$ we have for any $v \in H_0^1(\Omega)$

$$\langle \Pi_{\mathcal{G}} v \,,\, V \rangle = \langle v \,,\, V \rangle \qquad \text{for all } V \in \mathbb{V}(\mathcal{G}).$$

For simplicity, we abbreviate $\Pi_n := \Pi_{\mathcal{G}_n}$ for $n = 1, \ldots, N$. Hence, we can write equation (EG) as

$$\frac{1}{\tau_n} \langle U_n - \Pi_n U_{n-1} \,,\, V \rangle + \mathcal{B}(U_n \,,\, V) = \langle f_n \,,\, V \rangle \qquad \text{for all } V \in \mathbb{V}_n.$$

In this sense, equation (EG) for the solution U_n in the n-th time step may not resolve the whole information from $U_{n-1} \in \mathbb{V}_{n-1}$ but only its L^2-projection to the current finite element space $\mathbb{V}_n$, cf. Remark 3.2.2. Hence, we expect U_n to converge to $\Pi_n U_{n-1} \in \mathbb{V}_n$ rather than to $U_{n-1} \in \mathbb{V}_{n-1}$, which we confirm in Theorem 4.2.5 below.

This consideration suggests to use the L^2-projection $\Pi_n U_{n-1} \in \mathbb{V}_n$ for splitting the coarsen-time indicator into a *coarsen* and a *time indicator*. We cite the definition of those two indicators from (4.8): For $v, w \in H_0^1(\Omega)$, a triangulation $\mathcal{G}$ and an element $E \in \mathcal{G}$ the time respectively coarsen indicators are defined as

$$\mathcal{E}_\tau^2(v, w, \mathcal{G}) := 10 \left\| v - \Pi_{\mathcal{G}} w \right\|_\Omega^2,$$
$$\mathcal{E}_c^2(w, \mathcal{G}, E) := 10 \left\| \Pi_{\mathcal{G}} w - w \right\|_E^2.$$

Note that for nested finite element spaces $\mathbb{V}_{n-1} \subset \mathbb{V}_n$, we have $\Pi_n V = V$ for all $V \in \mathbb{V}_{n-1}$ and thus $\mathcal{E}_c^2(U_{n-1}, \mathcal{G}_n) = 0$.

In the following, we will examine how the time and coarsening indicator may be reduced individually. In light of the estimate

$$\mathcal{E}_{c\tau}^2(U_n, U_{n-1}) \leq \mathcal{E}_{\tau}^2(U_n, U_{n-1}, \mathcal{G}_n) + \mathcal{E}_c^2(U_{n-1}, \mathcal{G}_n), \qquad (4.24)$$

this is not only of interest in view of establishing the bounds (4.9), but also addresses the question of how to reduce the coarsen-time indicator $\mathcal{E}_{c\tau}^2(U_n, U_{n-1})$ to meet the bound (4.4d).

Coarsening error indicator

From the construction using the L^2-projection, we see that the coarsening error indicator $\mathcal{E}_c^2(U_{n-1}, \mathcal{G}_n) = 0$ in case of nested spaces $\mathbb{V}_{n-1} \subset \mathbb{V}_n$ and hence, refining the underlying mesh $\mathcal{G}_n$ towards $\mathcal{G}_{n-1}$ — in the same sense as in the treatment of the energy gain in Section 4.2.1 — eventually yields $\mathcal{G}_{n,K} \geq \mathcal{G}_{n-1}$ for some $K \geq 0$. This implies nested finite element spaces $\mathbb{V}_{n-1} \subset \mathbb{V}_{n,K}$ which in turn induces a zero coarsening error $\mathcal{E}_c^2(U_{n-1}, \mathcal{G}_{n,K}) = 0$. Thus, we employ a directed mesh refinement for reducing the coarsening error indicator, compare also Section 4.2.1.

Note that the coarsening error indicator, which we recall reads $\mathcal{E}_c^2(U_{n-1}, \mathcal{G}_n) = 10 \|\Pi_n U_{n-1} - U_{n-1}\|_{\Omega}^2$ exclusively involves the solution of the previous time step and is hence independent of any modifications of the current time step size τ_n. Thus, no coupling effects with changes of the time step size will occur for the coarsening error indicator.

Time error indicator

Opposed to that, the time error indicator given as $\mathcal{E}_{\tau}^2(U_n, U_{n-1}, \mathcal{G}_n)$ $= 10 \|U_n - \Pi_n U_{n-1}\|_{\Omega}^2$ does involve the current solution which hinges on both the current time step size and the current finite element space. Hence, both changes to τ_n and $\mathbb{V}_n$ may effect the time error indicator in general. However, the foregoing considerations suggest that decreasing the time step size τ_n is the appropriate way for reducing the time sub-indicator $\mathcal{E}_{\tau}^2(U_n, U_{n-1}, \mathcal{G}_n)$. Theorem 4.2.5, which is supported by the following lemma, confirms this intuition.

We emphasize that in ASTFEM, we do not exclusively rely on decreasing the time error indicators for temporal error control but rather use a special error control which also makes use of the uniform energy bound employed in error control B; details on this will be presented in the following Section 4.3.

Lemma 4.2.4
Let $\{v_n\}_{n\in\mathbb{N}} \subset H_0^1(\Omega)$ be a sequence which converges weakly in $L^2(\Omega)$ towards some $v \in H_0^1(\Omega)$, i.e., $v_n \rightharpoonup v$ in $L^2(\Omega)$. Moreover, let $\{v_n\}_{n\in\mathbb{N}}$ be bounded in $H_0^1(\Omega)$, i.e., $\|v_n\|_\Omega \leq C$ for all $n \in \mathbb{N}$. Then $\{v_n\}_{n\in\mathbb{N}}$ converges weakly to v also in $H_0^1(\Omega)$, i.e., $v_n \rightharpoonup v$ in $H_0^1(\Omega)$.

Proof. $\boxed{\text{A}}$ We first show that any weakly-H^1-convergent subsequence of $\{v_n\}_{n\in\mathbb{N}}$ converges to v. To that end, let $\{v_{n_k}\}_{k\in\mathbb{N}}$ be such a subsequence, i.e., $v_{n_k} \rightharpoonup \tilde{v}$ in $H_0^1(\Omega)$ for some $\tilde{v} \in H_0^1(\Omega)$. Since $\{v_{n_k}\}_{k\in\mathbb{N}}$ is a subsequence of $\{v_n\}_{n\in\mathbb{N}}$, we also have $v_{n_k} \rightharpoonup v$ in $L^2(\Omega)$ by assumption. As the embedding $H_0^1(\Omega) \subset L^2(\Omega)$ implies the dual embedding $(L^2(\Omega))' \subset H^{-1}(\Omega)$, we particularly have for $\varphi \in (L^2(\Omega))' \subset H^{-1}(\Omega)$

$$\langle v_{n_k}, \varphi \rangle_{H_0^1(\Omega) \times H^{-1}(\Omega)} = \langle v_{n_k}, \varphi \rangle_{L^2(\Omega) \times (L^2(\Omega))'}$$
$$\to \langle v, \varphi \rangle_{L^2(\Omega) \times (L^2(\Omega))'} = \langle v, \varphi \rangle_{H_0^1(\Omega) \times H^{-1}(\Omega)},$$

i.e., $v_{n_k} \rightharpoonup v$ and the uniqueness of the weak limit implies $\tilde{v} = v$.

$\boxed{\text{B}}$ With this established, we argue by contradiction and assume that $v_n \not\rightharpoonup v$ in $H_0^1(\Omega)$. Thus, there exists a subsequence $\{v_{n_l}\}_{l\in\mathbb{N}}$ which does not contain any subsequence weakly converging to v in $H_0^1(\Omega)$. Since $\{v_{n_l}\}_{l\in\mathbb{N}}$ is bounded in $H_0^1(\Omega)$ by assumption, the reflexivity of $H_0^1(\Omega)$ implies that there exists a weakly-H^1-convergent subsequence of $\{v_{n_l}\}_{l\in\mathbb{N}}$. From A we conclude that this subsequence converges to v; a contradiction. $\square$

Theorem 4.2.5 (Convergence of time error indicators for fixed FE space)
Let $U_{n-1} \in \mathbb{V}_{n-1}$ be given and denote by $\Pi_n U_{n-1} \in \mathbb{V}_n$ its L^2-projection to $\mathbb{V}_n$. Moreover, let $\{\tau_{n,k}\}_{k\in\mathbb{N}}$ be a sequence of positive time step sizes with $\tau_{n,k} \to 0$ as $k \to \infty$. Then the sequence of solutions $\{U_{n,k}\}_{k\in\mathbb{N}}$ to (EG$_k$) satisfies

$$\lim_{k\to\infty} \|U_{n,k} - \Pi_n U_{n-1}\|_\Omega = 0. \tag{4.25}$$

Proof. For convenience, we abbreviate

$$V_k := U_{n,k} - \Pi_n U_{n-1} \in \mathbb{V}_n$$

and note that V_k is an admissible test function for (EG_k). We thus get

$$\frac{1}{\tau_{n,k}} \|V_k\|^2 + \|V_k\|_\Omega^2 = \langle f_{n,k} , V_k \rangle - \mathcal{B}(\Pi_n U_{n-1} , V_k). \qquad (4.26)$$

We estimate the first term on the right hand side by Cauchy-Schwarz and Young's inequality to see

$$\langle f_{n,k} , V_k \rangle \le \frac{\tau_{n,k}}{2} \|f_{n,k}\|^2 + \frac{1}{2\tau_{n,k}} \|V_k\|^2$$

$$\le \frac{1}{2} \|f\|_{L^2(0,T;L^2(\Omega))}^2 + \frac{1}{2\tau_{n,k}} \|V_k\|^2 .$$

For the second term, we derive by continuity and Young's inequality

$$|\mathcal{B}(\Pi_n U_{n-1} , V_k)| \le \frac{\mathcal{C}_\mathcal{B}^2}{2} \|\Pi_n U_{n-1}\|_\Omega^2 + \frac{1}{2} \|V_k\|_\Omega^2$$

and substituting these estimates into (4.26) gives the following bound which is uniform in k:

$$\frac{1}{\tau_{n,k}} \|V_k\|^2 + \|V_k\|_\Omega^2 \le \|f\|_{L^2(0,T;L^2(\Omega))}^2 + \mathcal{C}_\mathcal{B}^2 \|\Pi_n U_{n-1}\|_\Omega^2 .$$

This implies that V_k is uniformly bounded in H^1 and $\|V_k\|_{L_2(\Omega)} \to 0$ as $k \to \infty$ since $\tau_{n,k} \to 0$. The strong convergence $V_k \to 0$ in L_2 then particularly implies weak convergence $V_k \rightharpoonup 0$ in L^2 and applying Lemma 4.2.4 reveals that the full sequence V_k converges weakly in H^1 to 0, i.e., $V_k \rightharpoonup 0$ as $k \to \infty$. Weak convergence of the full sequence in L_2 and H^1 in combination with (4.26) then finally yields

$$\frac{1}{\tau_{n,k}} \|V_k\|^2 + \|V_k\|_\Omega^2 = \langle f_{n,k} , V_k \rangle - \mathcal{B}(\Pi_n U_{n-1} , V_k) \longrightarrow 0$$

as $k \to \infty$ and especially the claim $\lim_{k\to\infty} \|U_{n,k} - \Pi_n U_{n-1}\|_\Omega = 0$ holds. $\square$

Motivated by this, we will reduce the time step size τ_n in order to reduce the time error indicator $\mathcal{E}_\tau^2(U_n, U_{n-1}, \mathcal{G}_n)$. In the basic Algorithm 4, we assume that we have an initial guess for the time step size $\tau_{n,0} > 0$ at our disposal.

Algorithm 4 Basic time error adaptation

Set $k := 0$;
Solve (EG_k) for $U_{n,k}$;
Compute $\mathcal{E}_\tau^2(U_{n,k}, U_{n-1}, \mathcal{G}_n)$;
while $\mathcal{E}_\tau^2(U_{n,k}, U_{n-1}, \mathcal{G}_n) > \text{TOL}_t$ **do**
    Set $k := k + 1$;
    Produce $\tau_{n,k}$ by decreasing $\tau_{n,k-1}$;
    Solve (EG_k) for $U_{n,k}$;
    Compute $\mathcal{E}_\tau^2(U_{n,k}, U_{n-1}, \mathcal{G}_n)$;
end while
Return $\tau_n := \tau_{n,k}$;

4.2.6 Coupling effects

In the preceeding Sections 4.2.1 through 4.2.5, we considered the reduction of one error indicator at a time and we also investigated how to ensure the energy gain criterion of (4.20). We found that in any given time step

- reducing the time step size is the correct means for reducing the consistency and the time error indicator, while

- refining the mesh is the appropriate means for reducing the coarsening and the space error indicator as well as for controlling the energy gain.

This implies that in general, error control in a single time step asks for both reducing the time step size and refining the mesh. In the following, we investigate any coupling effects arising from changing the time step size as well as the mesh. First, we analyze the indicators which solely depend on either the time step size or the mesh.

Uncoupled indicators

Again, we start with the initial error indicator $\mathcal{E}_0^2(u_0, \mathcal{G}_0)$ which is a pure data error. Recalling

$$\mathcal{E}_0^2(u_0, \mathcal{G}_0) = \|u_0 - \mathcal{P}_{\mathcal{G}_0} u_0\|^2,$$

we observe that this indicator exclusively depends on the given u_0 as well as the choice of the mesh $\mathcal{G}_0$; no dependence on any time step size is given.

Similarly, we see that also the consistency error

$$\mathcal{E}_f^2(f, t_{n-1}, \tau_n) = C_f \frac{1}{\tau_n} \int_{t_n}^{t_n + \tau_n} \|f(s) - f_n\|^2 \, ds$$

is a pure data error solely depending on the given right hand side f and the current time step size τ_n. In particular, $\mathcal{E}_f^2(f, t_{n-1}, \tau_n)$ does *not* depend on the choice of the mesh $\mathcal{G}_n$ underlying the finite element space $\mathbb{V}_n$.

The last of the uncoupled indicators is the coarsening indicator $\mathcal{E}_c^2(U_{n-1}, \mathcal{G}_n)$. This is the case since

$$\mathcal{E}_c^2(U_{n-1}, \mathcal{G}_n) = \|U_{n-1} - \Pi_n U_{n-1}\|_\Omega^2$$

exclusively depends on last time step's solution U_{n-1}, which can be regarded as fixed given data in the n-th time step, as well as the mesh $\mathcal{G}_n$. Above all, the coarsening error does *not* depend on the choice of the time step size τ_n.

Unfortunately, the remaining indicators do depend on both the time step size and the mesh.

Coupled indicators

Energy gain In Section 4.2.1 we saw that refining the mesh $\mathcal{G}_n$ is an appropriate means for establishing the energy gain criterion (4.20). However, from the definition of

$$\mathcal{E}_*^2(U_n, U_{n-1}, \tau_n, \mathcal{G}_n) = \frac{1}{\tau_n} \|U_n - \Pi_n U_{n-1}\|^2 + \|U_{n-1}\|_\Omega^2 - \|\Pi_n U_{n-1}\|_\Omega^2$$

we see that this quantity also depends on the time step size τ_n. Note that this dependency is not only given explicitly by the coefficient $\frac{1}{\tau_n}$ but also implicitly via the solution U_n, which also hinges on the employed time step size. Nevertheless, for any mesh $\mathcal{G}_{n,k} \geq \mathcal{G}_{n-1}$ we have nested finite element spaces $\mathbb{V}_{n-1} \subset \mathbb{V}_{n,k}$, which indeed implies

$$\mathcal{E}_*^2(U_{n,k}, U_{n-1}, \tau_{n,k}, \mathcal{G}_{n,k}) = \frac{1}{\tau_{n,k}} \|U_{n,k} - U_{n-1}\|^2 \geq 0$$

irrespective of the current time step size $\tau_{n,k}$, and hence provides the demanded energy gain criterion of (4.20), cf. Section 4.2.1.

Space indicator From Section 4.2.4 we recall that the space indicator $\mathcal{E}_{\mathcal{G}}^2(U_n, U_{n-1}, \tau_n, f_n, \mathcal{G}_n)$ is the error indicator of the associated stationary problem which reads: Find $u_n \in H_0^1(\Omega)$ such that

$$\mathcal{B}(u_n\,,\,v) + \frac{1}{\tau_n}\langle u_n\,,\,v\rangle = \langle f_n + \frac{1}{\tau_n}U_{n-1}\,,\,v\rangle \qquad \text{for all } v \in H_0^1(\Omega).$$

Since this elliptic problem involves the time step size τ_n as a coefficient, so does the related error indicator given by

$$\mathcal{E}_{\mathcal{G}}^2(U_n, U_{n-1}, \tau_n, f_n, \mathcal{G}_n)$$

$$= 15\,C_h \sum_{E \in \mathcal{G}_n} \left(h_E^2 \left\| f_n - \frac{U_n - U_{n-1}}{\tau_n} - \mathcal{L}U_n \right\|_{L^2(E)}^2 \right.$$

$$\left. + h_E \left\| [\![\mathbf{A}\nabla U_n]\!] \right\|_{L^2(\partial E \cap \Omega)}^2 \right).$$

Note that this indicator also depends on the averaged right hand side f_n introducing yet another dependency on the time step size τ_n.

On the other hand, we identified refinement of the mesh $\mathcal{G}_n$ as an appropriate means for reducing the space error indicator, whereas we do not know how changing the time step size τ_n affects $\mathcal{E}_{\mathcal{G}}^2(U_n, U_{n-1}, \tau_n, f_n, \mathcal{G}_n)$. In particular, having produced a mesh $\mathcal{G}_{n,k}$ in the k-th iteration such that $\mathcal{E}_{\mathcal{G}}^2(U_{n,k}, U_{n-1}, \tau_{n,k}, f_{n,k}, \mathcal{G}_{n,k})$ $\leq \text{TOL}_{\mathcal{G}}$ does not imply that such an error bound remains valid in the next iterative step when assuming that the time step size $\tau_{n,k+1} \neq \tau_{n,k}$ is changed. In fact, in iterative step $k+1$ we may well have $\mathcal{E}_{\mathcal{G}}^2(U_{n,k+1}, U_{n-1}, \tau_{n,k+1}, f_{n,k+1}, \mathcal{G}_{n,k+1}) > \text{TOL}_{\mathcal{G}}$.

Concluding, we state that whereas we know that for a fixed time step size the space indicator can be reduced by refining the mesh, this does not apply to situations of changing time step sizes.

Time indicator For the time indicator which we recall reads $\mathcal{E}_{\tau}^2(U_n, U_{n-1}, \mathcal{G}_n) = \|U_n - \Pi_n U_{n-1}\|^2$ we find a similar coupling. Whereas we found that for a fixed mesh $\mathcal{G}_n$ underlying the finite element space $\mathbb{V}_n$ the time indicator may be reduced by decreasing the time step size τ_n, we now observe that $\mathcal{E}_\tau^2(U_n, U_{n-1}, \mathcal{G}_n)$ in fact strongly depends on the current mesh $\mathcal{G}_n$. This is not only since it involves the L^2-projection to the related finite element space $\mathbb{V}_n$, but we also realize that the solution U_n hinges on the choice of the space $\mathbb{V}_n$ respectively its underlying mesh $\mathcal{G}_n$. In particular, having established the tolerance $\mathcal{E}_\tau^2(U_{n,k}, U_{n-1}, \mathcal{G}_{n,k}) \leq \texttt{TOL}_t$ in the k-th iteration does not exclude a violation of this bound in the next iterative step $k+1$, which we suppose involves a mesh $\mathcal{G}_{n,k+1} \neq \mathcal{G}_{n,k}$. We may rather have $\mathcal{E}_\tau^2(U_{n,k+1}, U_{n-1}, \mathcal{G}_{n,k+1}) > \texttt{TOL}_t$.

4.2.7 Establishing error control for coupled indicators

From the previous section we observe that the reduction of the space indicator via mesh refinement is indeed coupled to the reduction of the time indicator which involves decreasing the time step size. To resolve this issue and to guarantee that e.g., the time indicator will eventually meet its tolerance, we might demand that after a certain number $K \geq 0$ of iterations, the mesh $\mathcal{G}_{n,k}$ stays fixed for all $k \geq K$. However, by this approach, we abandon any control over the space indicator after the K-th iteration and hence cannot guarantee that the space tolerance will be fulfilled. Alternatively, we might demand that after the K-th iteration, the time step size stays fixed, i.e., $\tau_{n,k} = \tau_{n,K}$ for all $k \geq K$. Analogously to the above approach, this however implies that the time indicator may no longer be controlled after the K-th iteration and hence, the associated tolerance may not be satisfied. Nevertheless, we pursue this approach of eventually fixing the time step size $\tau_{n,k} = \tau_{n,K}$ for all $k \geq K$ and hence, error control for the space error is given as in the uncoupled case, cf. Sections 4.2.4 and 3.3. Thanks to Theorem 4.2.6 below, we will still be able to guarantee any demanded bound on the time error indicator. To allow for this, Theorem 4.2.6 states that (under certain conditions) reducing the time step size actually reduces the time error even if the corresponding meshes do change. This is an important generalization of Theorem 4.2.5, however, the proof follows the same lines and we give it for completeness.

Theorem 4.2.6 (Convergence of time error indicators)
Let $U_{n-1} \in \mathbb{V}_{n-1}$ be given and let $\{\mathbb{V}_{n,k}\}_{k\in\mathbb{N}}$ be an arbitrary sequence of finite element spaces. Further, denote by $\Pi_{n,k}U_{n-1} \in \mathbb{V}_{n,k}$ the L^2-projection of U_{n-1} into the space $\mathbb{V}_{n,k}$ and suppose that $\|\Pi_{n,k}U_{n-1}\|_\Omega^2$ is uniformly bounded, i.e., there exists a constant $C = C(n)$ which in particular is independent of k such that

$$\|\Pi_{n,k}U_{n-1}\|_\Omega^2 \le C \qquad \text{for all } k \in \mathbb{N}. \tag{4.27}$$

Moreover, let $\{\tau_{n,k}\}_{k\in\mathbb{N}}$ be a sequence of positive time step sizes with $\tau_{n,k} \to 0$ as $k \to \infty$. Then the sequence of solutions $\{U_{n,k}\}_{k\in\mathbb{N}}$ to (EG_k) satisfies

$$\lim_{k\to\infty} \|U_{n,k} - \Pi_{n,k}U_{n-1}\|_\Omega = 0.$$

Proof. The proof is very similar to the proof of Theorem 4.2.5. We have $\Pi_{n,k}U_{n-1} \in \mathbb{V}_{n,k}$ for all $k \in \mathbb{N}$. Hence, defining

$$V_k := U_{n,k} - \Pi_{n,k}U_{n-1}$$

we see that $V_k \in \mathbb{V}_{n,k}$ for all $k \in \mathbb{N}$ and therefore, $V_k \in \mathbb{V}_{n,k}$ is an admissible test function for (EG_k). Using $V_k \in \mathbb{V}_{n,k}$ as a test function in (EG_k) we derive

$$\frac{1}{\tau_{n,k}} \|V_k\|^2 + \|V_k\|_\Omega^2 = \langle f_{n,k}, V_k \rangle - \mathcal{B}(\Pi_{n,k}U_{n-1}, V_k).$$

and estimating the right hand side as in the proof of Theorem 4.2.5 gives the bound

$$\frac{1}{\tau_{n,k}} \|V_k\|^2 + \|V_k\|_\Omega^2 \le \|f\|_{L^2(0,T;L^2(\Omega))}^2 + \mathcal{C}_\mathcal{B}^2 \|\Pi_{n,k}U_{n-1}\|_\Omega^2.$$

By assumption (4.27) this bound is uniform in k. Again by the same arguments as in the proof of Theorem 4.2.5 this implies

$$\frac{1}{\tau_{n,k}} \|V_k\|^2 + \|V_k\|_\Omega^2 = \langle f_{n,k}, V_k \rangle - \mathcal{B}(\Pi_{n,k}U_{n-1}, V_k) \longrightarrow 0$$

as $k \to \infty$. In particular, we established the claim

$$\lim_{k\to\infty} \|U_{n,k} - \Pi_{n,k}U_{n-1}\|_\Omega = 0.$$

$\square$

Remark 4.2.7
A result for nested finite element spaces $\mathbb{V}_{n-1} \subset \mathbb{V}_{n,k}$ similar to the above theorem was first established by Chen and Feng in [14]. However, whereas Chen and Feng employ a continuous as well as a discrete auxiliary problem for establishing the convergence $\|U_{n,k} - \Pi_{n,k}U_{n-1}\|_\Omega \to 0$, the proof presented above only exploits the discrete equation (EG$_k$) by testing it appropriately. In this sense, the presented proof is more transparent which also allowed for a straightforward generalization to non-nested finite element spaces under the additional condition that $\|\Pi_{n,k}U_{n-1}\|_\Omega^2$ is uniformly bounded.

Remark 4.2.8 (Bounding $\|\Pi_{n,k}U_{n-1}\|_\Omega^2$)
Since the L^2-projection $\Pi_{n,k} : \mathbb{V}_{n-1} \to \mathbb{V}_{n,k}$ is *not* H^1-stable in the sense that

$$\|\Pi_{n,k}V\|_\Omega \le C \|V\|_\Omega \qquad \text{for all } V \in \mathbb{V}_{n-1}, \tag{4.28}$$

we may particularly *not* deduce the uniform bound $\|\Pi_{n,k}U_{n-1}\|_\Omega^2 \le \|U_{n-1}\|_\Omega^2$. However, under certain additional conditions regarding the mesh, Bramble, Pasciak, and Steinbach established the stability estimate (4.28) in [10]. Particularly, this H^1-stability is given for locally quasi-uniform meshes provided the volume of neighboring elements does not change too drastically.
We do not elaborate on these conditions but instead explicitly demand that $\|\Pi_{n,k}U_{n-1}\|_\Omega^2$ is bounded uniformly in k. In particular, this demand is satisfied whenever we have nested finite element spaces $\mathbb{V}_{n-1} \subset \mathbb{V}_{n,k}$ since this implies $\Pi_{n,k}U_{n-1} = U_{n-1}$ and thus $\|\Pi_{n,k}U_{n-1}\|_\Omega^2 = \|U_{n-1}\|_\Omega^2$. Moreover, we also mention that the desired bound for $\|\Pi_{n,k}U_{n-1}\|_\Omega$ is related to the energy gain criterion, since this implies $\|\Pi_{n,k}U_{n-1}\|_\Omega^2 \le \frac{1}{\tau_{n,k}} \|U_{n,k} - \Pi_{n,k}U_{n-1}\|^2 + \|U_{n-1}\|_\Omega^2$. However, we emphasize that this bound is *not* uniform in k as assumed in Theorem 4.2.6.

From Theorem 4.2.6 we see that indeed reducing the time step size is the appropriate action to be taken for reducing the time error indicator. We emphasize in particular that $\mathcal{E}_\tau^2(U_{n,k}, U_{n-1}, \mathcal{G}_{n,k}) \to 0$ as $\tau_{n,k} \to 0$ even if the corresponding finite element spaces $\mathbb{V}_{n,k}$ are changing provided $\|\Pi_{n,k}U_{n-1}\|_\Omega^2$ is uniformly bounded. This convergence of the time error indicators particularly implies that for each time step n, there exists a dedicated minimal time step size $\tau_n^* > 0$ guaranteeing that a given positive tolerance is reached. Especially, it is not necessary to further reduce the time step size once τ_n^* is reached (at least where the time error is concerned). This particularly implies that the elliptic problem of the current time step is fixed and hence, the spatial discretization error can be reduced by simply refining the mesh. In [14], Chen and Feng employed this kind of argumentation together with convergence for elliptic problems to guarantee that an adaptive algorithm (cf. Algorithm 9 on page 184), which employs error control in the style of error control A, terminates in each single time step. However, since the minimal time step size τ_n^* is only valid for the n-th time step, no global convergence can be achieved and particularly the question whether the final time is reached, stays unanswered. Opposed to this, in the following section we introduce a new way of error control, which especially builds on the *global* minimal time step size τ_* involved in error control B, to ensure overall termination and reaching the final time. In particular, we do not employ above Theorem 4.2.6 and its additional condition that $\|\Pi_{n,k} U_{n-1}\|_\Omega^2$ is uniformly bounded to establish convergence and termination of ASTFEM.

4.3 Error control in ASTFEM

In this section, we introduce the method of error control that will actually be employed in ASTFEM, cf. [27]. In the process, we heavily rely on the insights we gained during Sections 4.1 and 4.2. We consider the advantages and drawbacks of error control A respectively A' and error control B particularly in light of the means for enabling these controls presented in Section 4.2.

4.3.1 Comparison of error control strategies

Above all, we elaborate on the coarsen-time indicator and especially the time sub-indicator since this is treated alternatively in error control B. More precisely, we compare error control A' and error control B in respect of bounding the overall temporal discretization error $\sum_{n=1}^{N} \tau_n \mathcal{E}_\tau^2(U_n, U_{n-1}, \mathcal{G}_n)$. We start by revisiting error control A' for the time error indicator, which demands

$$\mathcal{E}_\tau^2(U_n, U_{n-1}, \mathcal{G}_n) \leq \text{TOL}_t \qquad (4.29)$$

for each time step n. In Section 4.2.7, we saw that the time error indicator is reduced by reducing the time step size. In particular, we deduced from Theorem 4.2.6

$$\mathcal{E}_\tau^2(U_{n,k}, U_{n-1}, \mathcal{G}_{n,k}) \to 0 \qquad \text{as } \tau_{n,k} \to 0$$

provided $\|\Pi_{n,k} U_{n-1}\|_\Omega^2$ is uniformly bounded. We conclude that for any $\text{TOL}_t > 0$ there is a time step size τ_n^* such that

$$\mathcal{E}_\tau^2(U_{n,k}, U_{n-1}, \mathcal{G}_{n,k}) \leq \text{TOL}_t \qquad \text{for all } \tau_{n,k} \leq \tau_n^*.$$

Once again, we emphasize that this bound holds irrespectively of possible mesh modifications. We hereby established a minimal time step size τ_n^* for every time step n which guarantees that the time error control (4.29) will be satisfied once τ_n^* is reached. In other words, control of the time error does not require to reduce the time step size below τ_n^*. Hence, time error control will not contradict the termination of a single time step, since τ_n^* may be reached with a finite number of reductions of τ_n.

Note however, that this minimal time step size τ_n^* is only valid for the n-th time step and each time step has its own dedicated τ_n^*. Particularly for two different time steps $n_1 \neq n_2$ we generally have $\tau_{n_1}^* \neq \tau_{n_2}^*$. We assume that each time step employs its minimal time step size τ_n^*. Lacking a correlation between those dedicated minimal time step sizes τ_n^*, we have no control of the sum of all time step sizes. In particular, it is not clear whether T will be reached in finitely many steps and we may well have $N = \infty$. Moreover, premature convergence $\sum_{n=1}^\infty \tau_n^* = T' < T$ may appear, preventing any algorithm using the time step sizes $\{\tau_n^*\}_{n\geq 0}$ from reaching the desired final time T. This is the big drawback of error control A respectively A'.

On the other hand, error control B provides a *global* minimal time step size $\tau_* > 0$ which guarantees

$$\sum_{n=1}^N \tau_n \mathcal{E}_\tau^2(U_n, U_{n-1}, \mathcal{G}_n) \leq \text{TOL}_*$$

provided all time step sizes are sufficiently small, i. e., $\tau_n \leq \tau_*$ for all $n = 1, \ldots, N$. Hence, choosing $\tau_n = \tau_*$ for all $n = 1, \ldots, N$, the time error is controlled and still, after $N := \lceil \frac{T}{\tau_*} \rceil$ time steps (where $\lceil \cdot \rceil$ denotes the ceiling function), the final time is reached since $\sum_{n=1}^{N} \tau_n = N\tau_* \geq T$. Opposed to that, the big drawback of error control B is the massive constraint that all time step sizes must be as small as τ_*, i. e., $\tau_n \leq \tau_*$. This a priori bound is highly unwanted in an adaptive algorithm and contradicts every notion of (temporal) adaptivity.

Concluding, we state that exclusively employing error control A' may not guarantee that the final time T is reached whereas solely employing error control B contradicts adaptivity. Hence, we wish to derive an error control strategy that combines the advantages of both strategies while circumventing their respective drawbacks.

4.3.2 Joint error control

To do so, we mainly focus on the temporal discretization error and the related coarsen-time error before including the remaining error indicators into our considerations. We assume that for $n = 1, \ldots, N$, we have at our disposal time step sizes τ_n, finite element spaces $\mathbb{V}_n$ and corresponding solutions U_n of (EG). Since we also want to employ error control B, we further assume that the energy gain criterion $\mathcal{E}_*^2(U_n, U_{n-1}, \tau_n, \mathcal{G}_n) \geq 0$ is satisfied for all $n = 1, \ldots, N$.

The basic idea behind the joint error control is to primarily employ error control A, which is more adaptive, and to resort to error control B if in the process the time step size is getting too small which might contradict reaching the final time. In this case, we observe that error control B can be used to bound the temporal error arising from time steps with "small time step size" even if the respective time error indicators do not meet their tolerances. To be more precise, we split the set of all time step indices into disjoint sets $\{1, \ldots, N\} = \mathcal{A} \cup \mathcal{B}$, $\mathcal{A} \cap \mathcal{B} = \emptyset$, with $\mathcal{A}$ containing the indices of all time steps with "big" time step size, i. e., $\tau_n > \tau_*$, while $\mathcal{B}$ contains the indices of all time steps with "small" time step size,

i. e., $\tau_n \leq \tau_*$;

$$\mathcal{A} := \left\{ n \in \{1, \ldots, N\} \mid \tau_n > \tau_* \right\} \quad \text{and}$$
$$\mathcal{B} := \left\{ n \in \{1, \ldots, N\} \mid \tau_n \leq \tau_* \right\}.$$

Hence, we may split the temporal error into a share corresponding to big time steps and a share corresponding to small time steps, or, more precisely

$$\sum_{n=1}^{N} \tau_n \mathcal{E}_\tau^2(U_n, U_{n-1}, \mathcal{G}_n) = \sum_{n \in \mathcal{A}} \tau_n \mathcal{E}_\tau^2(U_n, U_{n-1}, \mathcal{G}_n)$$
$$+ \sum_{n \in \mathcal{B}} \tau_n \mathcal{E}_\tau^2(U_n, U_{n-1}, \mathcal{G}_n).$$

In particular, for small time steps $n \in \mathcal{B}$, we may employ error control B and we thereby deduce

$$\sum_{n \in \mathcal{B}} \tau_n \mathcal{E}_\tau^2(U_n, U_{n-1}, \mathcal{G}_n) \leq \tau_* \sum_{n=1}^{N} \mathcal{E}_\tau^2(U_n, U_{n-1}, \mathcal{G}_n) \leq \texttt{TOL}_*, \quad (4.30)$$

as we assumed that the energy gain criterion $\mathcal{E}_*^2(U_n, U_{n-1}, \tau_n, \mathcal{G}_n) \geq 0$ is satisfied for all $n = 1, \ldots, N$. We emphasize that in this case, *no control* of the time error indicator $\mathcal{E}_\tau$ has to be provided. For time steps $n \in \mathcal{A}$ on the other hand, we may not employ error control B as the respective time step sizes are too big. For this reason, we assume that the conditions of error control A are satisfied for $n \in \mathcal{A}$. Focusing on the coarsen-time error indicator which is involved in error control A, we again split

$$\sum_{n=1}^{N} \tau_n \mathcal{E}_{c\tau}^2(U_n, U_{n-1}) = \sum_{n \in \mathcal{A}} \tau_n \mathcal{E}_{c\tau}^2(U_n, U_{n-1}) + \sum_{n \in \mathcal{B}} \tau_n \mathcal{E}_{c\tau}^2(U_n, U_{n-1})$$

and for $n \in \mathcal{B}$, we estimate

$$\leq \sum_{n \in \mathcal{A}} \tau_n \mathcal{E}_{c\tau}^2(U_n, U_{n-1}) + \sum_{n \in \mathcal{B}} \tau_n \mathcal{E}_\tau^2(U_n, U_{n-1}, \mathcal{G}_n)$$
$$+ \sum_{n \in \mathcal{B}} \tau_n \mathcal{E}_c^2(U_{n-1}, \mathcal{G}_n).$$

By (4.30) and applying error control A to the terms with $n \in \mathcal{A}$, we obtain

$$\leq \text{TOL}_* + \sum_{n \in \mathcal{A}} \tau_n \text{TOL}_{\text{ct}} + \sum_{n \in \mathcal{B}} \tau_n \mathcal{E}_c^2(U_{n-1}, \mathcal{G}_n). \tag{4.31}$$

Hence, we notice that for time steps $n \in \mathcal{B}$, we additionally need to control the term $\sum_{n \in \mathcal{B}} \tau_n \mathcal{E}_c^2(U_{n-1}, \mathcal{G}_n)$ which we accomplish by demanding a bound

$$\mathcal{E}_c^2(U_{n-1}, \mathcal{G}_n) \leq \text{TOL}_c \qquad \text{for all } n \in \mathcal{B}$$

in fashion of error control A'. Assuming such a bound in (4.31) gives

$$\sum_{n=1}^{N} \tau_n \mathcal{E}_{c\tau}^2(U_n, U_{n-1}) \leq \text{TOL}_* + \sum_{n \in \mathcal{A}} \tau_n \text{TOL}_{\text{ct}} + \sum_{n \in \mathcal{B}} \tau_n \text{TOL}_c. \tag{4.32}$$

We hereby produced a bound on the coarsen-time error by imposing conditions in the style of error control A in big time steps $n \in \mathcal{A}$ while using error control B only in time steps which are small anyway, i.e., $n \in \mathcal{B}$.

We will use this philosophy for controlling the coarsen-time error indicator in the error control of ASTFEM. The remaining initial, consistency and space error indicators, for which no alternate error control in the fashion of error control B is available, will be dealt with by error control A. For the error control strategy actually employed in ASTFEM, we wish to distinctly bring out the differentiation between the two types of time steps $n \in \mathcal{A}$ and $n \in \mathcal{B}$. We therefore recall the a posteriori error estimate of Corollary 3.4.7

$$\|u - U\|_{\mathbb{W}(0,T)}^2 \leq \mathcal{E}_0^2(u_0, \mathcal{G}_0) + \sum_{n=1}^{N} \tau_n \left(\mathcal{E}_{\mathcal{G}}^2(U_n, U_{n-1}, \tau_n, f_n, \mathcal{G}_n) \right.$$
$$\left. + \mathcal{E}_{c\tau}^2(U_n, U_{n-1}) + \mathcal{E}_f^2(f, t_{n-1}, \tau_n) \right)$$

and rearrange the sums on the right hand side as

$$\|u - U\|^2_{W(0,T)} \leq \mathcal{E}_0^2(u_0, \mathcal{G}_0) + \sum_{n=1}^{N} \tau_n \mathcal{E}_f^2(f, t_{n-1}, \tau_n)$$
$$+ \sum_{n \in \mathcal{A}} \tau_n \big(\mathcal{E}_{\mathcal{G}}^2(U_n, U_{n-1}, \tau_n, f_n, \mathcal{G}_n) + \mathcal{E}_{c\tau}^2(U_n, U_{n-1}) \big)$$
$$+ \sum_{n \in \mathcal{B}} \tau_n \big(\mathcal{E}_{\mathcal{G}}^2(U_n, U_{n-1}, \tau_n, f_n, \mathcal{G}_n) + \mathcal{E}_c^2(U_{n-1}, \mathcal{G}_n) \big)$$
$$+ \sum_{n \in \mathcal{B}} \tau_n \mathcal{E}_\tau^2(U_n, U_{n-1}, \mathcal{G}_n)$$

where we may estimate the last term as in (4.30). With this in mind, we formulate and conclude error control C, which will be used in ASTFEM, cf. [27].

Error control C

Let TOL_0, TOL_f, TOL_*, and TOL_S be positive tolerances. Assume that

(1) the initial error is controlled via error control A, i. e.,

$$\mathcal{E}_0^2(u_0, \mathcal{G}_0) \leq TOL_0.$$

Assume further that for all $n = 1, \ldots, N$

(2) the energy gain condition

$$\mathcal{E}_*^2(U_n, U_{n-1}, \tau_n, \mathcal{G}_n) \geq 0$$

is satisfied;

(3) the consistency error indicator is controlled via error control A, i. e.,

$$\mathcal{E}_f^2(f, t_{n-1}, \tau_n) \leq TOL_f.$$

Moreover, suppose that for $n \in \mathcal{A}$

(4) the space and coarsen-time error indicators are controlled in the fashion of error control A via

$$\mathcal{E}_{\mathcal{G}}^2(U_n, U_{n-1}, \tau_n, f_n, \mathcal{G}_n) + \mathcal{E}_{c\tau}^2(U_n, U_{n-1}) \leq TOL_S.$$

Finally, assume that for $n \in \mathcal{B}$

(5) the space and coarsen error indicators are controlled in the fashion of error control A' via

$$\mathcal{E}_{\mathcal{G}}^2(U_n, U_{n-1}, \tau_n, f_n, \mathcal{G}_n) + \mathcal{E}_c^2(U_{n-1}, \mathcal{G}_n) \leq TOL_S.$$

These assumptions provide the error control

$$\|u - U\|_{W(0,T)}^2 \leq TOL_0 + TOL_* + \sum_{n=1}^{N} \tau_n (TOL_f + TOL_S).$$

Beyond distinguishing between small and big time steps and controlling the time discretization error of small time steps by taking advantage of the uniform energy bound (cf. Proposition 4.1.3), error control C also considers a pooled tolerance TOL_S. This unites the space and coarsen-time tolerances of error control A and is employed to enhance the flexibility of ASTFEM. In particular, considering a pooled tolerance allows for a large space error indicator in case of a small coarsen-time indicator and vice versa; see also the comparison of ASTFEM and a standard adaptive algorithm in Section 5.2.

4.4 Structure of ASTFEM

In this section, we implement error control C to state the adaptive algorithm ASTFEM (see page 155). To that end, we add details to the basic time stepping scheme introduced in Algorithm 2. ASTFEM then consists of several abstract modules which we do not present explicitly in this section. Rather, we characterize those modules by stating their essential properties, which we will exploit in Section 4.5 for showing that ASTFEM terminates while reaching any prescribed positive tolerance.

From the preceding sections, we recall that error control hinges on the employed time step sizes and finite element spaces. We hence elaborate on the generation of both a suitable time step size τ_n and a mesh $\mathcal{G}_n$ underlying the finite element space $\mathbb{V}_n$ in any time step n. By the term "suitable", to be more precise, we understand that

(1) the final time T shall be reached while

(2) the overall tolerance $\|u - U\|_{\mathbb{W}(0,T)}^2 \leq$ TOL is satisfied.

The line references throughout this section always refer to AST-FEM on page 155 unless stated otherwise.

4.4.1 Preliminaries

As a very first action in ASTFEM before actually entering the time stepping loop of Algorithm 2, we need to initialize objects. In particular, we provide an initial triangulation $\mathcal{G}_{\text{init}}$, an initial time step size τ_0 and we set $t_0 := 0$. Note however, that the initial mesh may be chosen arbitrarily and not necessarily provides the desired control for the initial error. On the contrary, in general we have $\mathcal{E}_0^2(u_0, \mathcal{G}_{\text{init}}) >$ TOL$_0$. Hence, before entering the time stepping, we first have to provide appropriate control of the initial error. This means, a mesh $\mathcal{G}_0$ has to be generated such that $\mathcal{E}_0^2(u_0, \mathcal{G}_0) \leq$ TOL$_0$. The mesh $\mathcal{G}_0$ gives rise to the finite element space $\mathbb{V}_0 := \mathbb{V}(\mathcal{G}_0)$ and we use the corresponding approximation $\mathcal{P}_{\mathcal{G}_0} u_0 \in \mathbb{V}_0$ of u_0 to define the first member of the sequence of solutions $\{U_n\}_{n \in \mathbb{N}}$, cf. Remark 4.1.1.

We assign the actual generation of such a mesh $\mathcal{G}_0$ to an abstract module ADAPT_INIT, which takes as an input the initial value u_0 as well as the initial mesh $\mathcal{G}_{\text{init}}$. In general, ADAPT_INIT possesses the properties stated below which make sure that the initial data are resolved sufficiently and the module terminates. Since the module ADAPT_INIT is only responsible for data approximation, we legitimately assume it as a black box, see Remark 4.2.1. Naturally, the first action in ASTFEM is a call of ADAPT_INIT, see line 2.

Properties 4.4.1 (ADAPT_INIT)
Let $v \in L^2(\Omega)$ and suppose $\mathcal{G}$ is a conforming triangulation of Ω. The module

$$(V, \mathcal{G}_*) = \text{ADAPT\_INIT}(v, \mathcal{G})$$

generates a refinement $\mathcal{G}_* \geq \mathcal{G}$ of $\mathcal{G}$ such that the approximation $\mathcal{P}_{\mathcal{G}_*} v \in \mathbb{V}(\mathcal{G}_*)$ satisfies

$$\mathcal{E}_0^2(v, \mathcal{G}_*) = \|v - \mathcal{P}_{\mathcal{G}_*} v\|^2 \leq \text{TOL}_0.$$

The module ADAPT_INIT$(v, \mathcal{G})$ terminates and returns the approximation $V := \mathcal{P}_{\mathcal{G}_*} v$ and the generated grid $\mathcal{G}_*$.

Remark 4.4.2 (Initial mesh $\mathcal{G}_{\text{init}}$)
We will use the initial mesh $\mathcal{G}_{\text{init}}$ as a basis for all meshes $\mathcal{G}_n$ produced throughout the algorithm ASTFEM in the sense that $\mathcal{G}_n \geq \mathcal{G}_{\text{init}}$. This clearly motivates to choose $\mathcal{G}_{\text{init}}$ as coarse as possible in order to provide maximal flexibility to the meshes $\mathcal{G}_n$. Hence, to allow for such a coarse initial mesh, we should not demand the initial tolerance to be satisfied on $\mathcal{G}_{\text{init}}$ but rather employ the above approach of generating a $\mathcal{G}_0$ satisfying the initial tolerance.

Moreover, before entering the time stepping, for a given $\text{TOL}_* > 0$ we compute the minimal time step size $\tau_* = \tau_*(U_0, f) > 0$ which is defined in (4.22) in error control B. We stress that τ_* exclusively depends on the discrete initial value U_0, the right hand side f, and in case of a general non-symmetric bilinear form $\mathcal{B}$ also on $\mathcal{B}$ (via the constant $\mathcal{C}_{\mathcal{B}_N}$) and the domain Ω (via the Poincaré constant $\mathcal{C}_\mathcal{P}$).

After those preliminaries, we enter the actual time stepping.

4.4.2 A single time step

We consider an arbitrary time step n (implemented in lines 5–14) for which we aim at computing a solution $U_n \in \mathbb{V}_n$ which complies with the demands of error control C. Recalling Section 4.2, this hinges on the choice of the dedicated time step size τ_n and mesh $\mathcal{G}_n$ and we thus focus on the generation of suitable τ_n and $\mathcal{G}_n$. As suggested in Section 4.2, this is done in an iterative procedure giving rise to sequences $\{\tau_{n,k}\}_{k \geq 0}$ and $\{\mathcal{G}_{n,k}\}_{k \geq 0}$.

Coarsening

As a very first action in each time step, we deal with providing an appropriate initial guess $\mathcal{G}_{n,0}$ for a mesh of the current time step. To that end, we produce a coarsening of the grid $\mathcal{G}_{n-1}$ of the previous time step. We point out that also this procedure does not involve computing the solution U_n of the current time step. The coarsening shall be accomplished by a black box module MARK_COARSEN (called in line 6) which produces the grid $\mathcal{G}_{n,0}$ as a coarsening of $\mathcal{G}_{n-1}$. More generally again, we introduce the corresponding module MARK_COARSEN, which might additionally consider information about the previous solution U_{n-1} to coarsen the mesh $\mathcal{G}_{n-1}$ cleverly, cf. Remark 4.4.4 below. At this point, we already anticipate that we only allow for refinements in the remaining time step.

Properties 4.4.3 (MARK_COARSEN)
The module MARK_COARSEN expects as input a conforming triangulation $\mathcal{G}$ and a finite element function $V \in \mathbb{V}(\mathcal{G})$. It generates a conforming coarsening $\mathcal{G}_* \leq \mathcal{G}$ of $\mathcal{G}$. To perform this operation, MARK_COARSEN may use information from V. The module terminates and returns the coarsened mesh $\mathcal{G}_*$,

$$\mathcal{G}_* = \text{MARK\_COARSEN}(V, \mathcal{G}).$$

Remark 4.4.4 (Purposeful coarsening)
The module MARK_COARSEN does not provide any error control at all and is thus very flexible in the way the coarse grid $\mathcal{G}_{n,0} = \text{MARK\_COARSEN}(U_{n-1}, \mathcal{G}_{n-1})$ is created from $\mathcal{G}_{n-1}$. However, assuming that the exact solution u does not change too much between two consecutive time steps, it seems appropriate to coarsen $\mathcal{G}_{n-1}$ only in such a way that U_{n-1} may still be "represented well" in the finite element space associated to the coarsened mesh. In this way, the coarsening indeed seems to be a good initial guess for a mesh in the current time step and we expect that this keeps low the number of iterations needed to generate the ultimate mesh $\mathcal{G}_n$. Moreover, such a coarsening is also reasonable in view of the lower bound on the spatial discretization error: As $U_{n-1} \in \mathbb{V}(\mathcal{G}_{n-1})$ enters the oscillation, a harsh coarsening unrelated to U_{n-1} may lead to a massive overestimation of the spatial discretization error, cf. Section 3.4.2. The actual module MARK_COARSEN used in the computations of Chapters 5 and 6 produces a coarsened mesh $\mathcal{G}_{n,0}$ such that $\left\|\left\|U_{n-1} - \mathcal{I}_{\mathcal{G}_{n,0}} U_{n-1}\right\|\right\|_\Omega^2$ is small, where $\mathcal{I}_{\mathcal{G}_{n,0}} : \mathbb{V}_{n-1} \to \mathbb{V}(\mathcal{G}_{n,0})$ denotes the Lagrange interpolation operator. Details on this particular coarsening strategy can be found in Section 5.1.2.

Remark 4.4.5 (Different coarsening strategies)
A wide-spread method in adaptive algorithms for parabolic problems is to delay the coarsening until the very end of each time step and to solve the current problem again on the coarsened mesh, cf. [14]. In Algorithm 9 on page 184 we state such an algorithm for comparison and we comment on the differences in Section 5.2.

Consistency error

Next, we consider the consistency error $\mathcal{E}_f^2(f, t_{n-1}, \tau_n)$ as it is a pure data error. Hence, producing a time step size τ_n such that the respective consistency tolerance is met, can be accomplished without computation of the solution U_n for the n-th time step.

From the considerations regarding the control of the consistency error indicator in Section 4.2.3, we see that we may produce a time step size $\tau_{n,K}$ such that $\mathcal{E}_f^2(f, t_{n-1}, \tau_{n,K}) \leq \mathrm{TOL}_f$ by successively decreasing a $\tau_{n,0}$ giving rise to a a sequence $\{\tau_{n,k}\}_{k \geq 0}$. To trigger this iterative process, we usually employ the time step size of the previous time step and set $\tau_{n,0} := \tau_{n-1}$.

More abstractly, we hand the task of ensuring the consistency tolerance to the module CONSISTENCY, which takes as arguments the right hand side f, the time node t_{n-1} and an initial guess $\tau_{n,0}$ for the current time step size. In the following, we state important properties of CONSISTENCY. In particular, we anticipate that CONSISTENCY complies with a minimal time step size $\tau_f > 0$ (cf. Remark 4.4.7), which we will determine in Lemma 4.6.1 using knowledge about the reduction of the consistency error from Lemma 4.2.2. A concrete realization of the module CONSISTENCY is given in Algorithm 6 on page 160.

Properties 4.4.6 (CONSISTENCY)
Let $f \in H^1(0, T; L^2(\Omega))$ and consider a time $t \in (0, T)$ and a time step size $\tau^{\mathrm{in}} \in (0, T)$. Moreover, let $\tau_f > 0$ be a given global parameter. Then the module

$$(\bar{f}, \tau) = \mathrm{CONSISTENCY}(f, t, \tau^{\mathrm{in}})$$

generates a time step size $\tau \in (0, T)$ with either

$$(1) \quad \min\{\tau_f, \tau^{\mathrm{in}}\} \leq \tau \leq T - t \qquad \text{or} \qquad (2) \quad \tau = T - t.$$

For the generated τ it holds $\mathcal{E}_f^2(f, t, \tau) \leq \mathrm{TOL}_f$. The module CONSISTENCY terminates and returns the mean value

$$\bar{f} := \frac{1}{\tau} \int_t^{t+\tau} f(s)\, ds \in L^2(\Omega)$$

as well as the modified time step size τ.

Remark 4.4.7 (Properties of returned τ)
The parameter $\tau_f > 0$ involved in above properties should be regarded as a global minimal time step size where the module CONSISTENCY is concerned. This is reflected in the properties of the returned time step size τ, which may not deceed τ_f with two exceptions: Firstly, (1) allows that if the input time step size τ^{in} is smaller that τ_f, the returned time step size τ may be as small as τ^{in}. Secondly, (2) allows that if the remaining time $T - t$ is smaller than both τ_f and τ^{in}, the returned time step size τ may equal $T - t$. Moreover, the returned time step size always satisfies $\tau \leq T - t$ and we point out that (1) is the main property of the returned τ whereas (2) only sees to it that the final time T is not exceeded.

Remark 4.4.8 (Increasing the time step size)
Note that whereas an initial guess $\tau_{n,0}$ for the time step size in the n-th time step which evokes too big a consistency error, i.e., $\mathcal{E}_f^2(f, t_{n-1}, \tau_{n,0}) > \text{TOL}_f$, requires to iteratively produce a time step size $\tau_{n,k} < \tau_{n,0}$ in order to satisfy the tolerance, an initial guess $\tau_{n,0}$ with $\mathcal{E}_f^2(f, t_{n-1}, \tau_{n,0}) \ll \text{TOL}_f$ does not *need* any further action from an error control point of view. However, employing a time step size τ_n which distinctly deceeds the demanded tolerance seems like wasting resources. Hence, in such cases, we will increase the time step size to finally obtain a $\tau_{n,K}$ such that

$$\delta\text{TOL}_f \leq \mathcal{E}_f^2(f, t_{n-1}, \tau_{n,K}) \leq \text{TOL}_f$$

for some parameter $\delta \in (0, 1)$, compare the realization of the module CONSISTENCY given in Algorithm 6 on page 160. In this sense, the generated time step size is adapted well to the right hand side f. However, we point out that this increasing of the time step size is not part of above essential properties of CONSISTENCY.

The module CONSISTENCY is called as the second action in each time step in line 7. To ensure that the consistency tolerance stays valid once CONSISTENCY$(f, t_{n-1}, \tau_{n,0})$ established $\mathcal{E}_f^2(f, t_{n-1}, \tau_{n,K})$ $\leq \text{TOL}_f$ for some $\tau_{n,K}$, we do not allow for increasing the time step size in the sequel of the n-th time step, i.e., $\tau_{n,k} \leq \tau_{n,K}$ for all $k \geq K$. By Lemma 4.2.2 this implies $\mathcal{E}_f^2(f, t_{n-1}, \tau_{n,k}) \leq \mathcal{E}_f^2(f, t_{n-1}, \tau_{n,K}) \leq \text{TOL}_f$ for all $k \geq K$.

Space and coarsen-time error

Until now, we only considered the uncoupled initial and consistency error indicators, which are independent of the solution U_n in the n-th time step. Before we can evaluate the a posteriori space and coarsen-time error indicators, it is necessary to compute the current time step's solution $U_{n,k}$ corresponding to the time step size $\tau_{n,k}$ and finite element space $\mathbb{V}_{n,k}$ generated in the course of the current time step. In order to compute this solution, we employ the black box module SOLVE which we state abstractly in the following.

Properties 4.4.9 (SOLVE)
Let $v, \bar{f} \in L^2(\Omega)$ and consider a time step size $\tau > 0$ and a conforming triangulation $\mathcal{G}$ of Ω. Then, the module

$$U := \texttt{SOLVE}(v, \bar{f}, \tau, \mathcal{G})$$

terminates and computes and returns the solution $U \in \mathbb{V}(\mathcal{G})$ of

$$\frac{1}{\tau}\langle U - v, V\rangle + \mathcal{B}(U, V) = \langle \bar{f}, V\rangle \qquad \text{for all } V \in \mathbb{V}(\mathcal{G}).$$

Having at hand this solution, we check whether the error control conditions from error control C are satisfied. In particular, we differentiate between big and small current time step sizes $\tau_{n,k} > \tau_*$ respectively $\tau_{n,k} \leq \tau_*$. In order to fulfill the conditions of error control C, we may have to reduce the time step size and/or refine the mesh. This all — including computing the current solution via SOLVE — is accomplished by the module ST_ADAPTATION which in this sense is the main module of the whole algorithm. It acts on the current time step size $\tau_{n,k}$, the corresponding mean $f_{n,k}$ of the right hand side and the current mesh $\mathcal{G}_{n,k}$. Moreover, as ST_ADAPTATION particularly solves the current elliptic problem to update the error estimates, it also employs the solution U_{n-1} from last time step. As a last argument, ST_ADAPTATION expects the mesh $\mathcal{G}_{n-1}$ of the previous time step since dealing with the coarsening error involves directed mesh refinements towards $\mathcal{G}_{n-1}$, cf. Section 4.2.5. More abstractly, ST_ADAPTATION possesses the following essential properties.

Properties 4.4.10 (ST_ADAPTATION)
Let $\mathcal{G}^{\text{in}}$ and $\mathcal{G}^{\text{old}}$ be conforming triangulations of Ω. Moreover, let

$v \in \mathbb{V}(\mathcal{G}^{\text{old}})$, $\bar{f}^{\text{in}} \in L^2(\Omega)$ and $t, \tau^{\text{in}} \in (0, T)$ such that $t + \tau^{\text{in}} \in (0, T)$. Then, the module

$$(U, \tau, \bar{f}, \mathcal{G}) = \text{ST\_ADAPTATION}(v, t, \tau^{\text{in}}, \bar{f}^{\text{in}}, \mathcal{G}^{\text{in}}, \mathcal{G}^{\text{old}})$$

generates a time step size τ satisfying either

$$(1) \quad \tau_* \leq \tau \leq \tau^{\text{in}} \qquad \text{or} \qquad (2) \quad \tau = \tau^{\text{in}},$$

and if the time step size is changed, i. e., $\tau \neq \tau^{\text{in}}$, $\bar{f} := \frac{1}{\tau} \int_t^{t+\tau} f(s) ds$ is updated, otherwise $\bar{f} := \bar{f}^{\text{in}}$. Moreover, it produces a refinement $\mathcal{G} \geq \mathcal{G}^{\text{in}}$ of $\mathcal{G}^{\text{in}}$ and computes the solution $U \in \mathbb{V}(\mathcal{G})$ of

$$\frac{1}{\tau} \langle U - v, V \rangle + \mathcal{B}(U, V) = \langle \bar{f}, V \rangle \qquad \text{for all } V \in \mathbb{V}(\mathcal{G}).$$

In case of a big time step, i. e., $\tau > \tau_*$, it holds

$$\mathcal{E}_{\mathcal{G}}^2(U, v, \tau, \bar{f}, \mathcal{G}) + \mathcal{E}_{c\tau}^2(U, v) \leq \text{TOL}_{\text{s}}$$

whereas in case of a small time step, i. e., $\tau \leq \tau_*$, we have

$$\mathcal{E}_{\mathcal{G}}^2(U, v, \tau, \bar{f}, \mathcal{G}) + \mathcal{E}_c^2(U, v) \leq \text{TOL}_{\text{s}}.$$

The module ST_ADAPTATION terminates and returns the solution $U \in \mathbb{V}(\mathcal{G})$, the generated time step size τ with corresponding mean value $\bar{f}$ as well as the generated mesh $\mathcal{G}$.

Remark 4.4.11
The arguments of ST_ADAPTATION bearing a superscript $\cdot^{\text{in}}$ are changed inside the module. The other arguments v, t, and $\mathcal{G}^{\text{old}}$ are not modified and only considered as data. In particular, the time argument t is only needed for the computation of $\bar{f}$.

ST_ADAPTATION is first called in line 8; a realization of the module can be found in Algorithm 7 on page 163.

Energy gain

The last condition of error control C is that the energy gain criterion $\mathcal{E}_*^2(U_n, U_{n-1}, \tau_n, \mathcal{G}_n) \geq 0$ is met in each time step. We emphasize that this condition must be fulfilled in any time step regardless of whether the minimal time step size is employed in the particular step or not. From Section 4.2.1 we recall that a suitable means for establishing the energy gain criterion is to refine the current mesh "towards" the "target" mesh $\mathcal{G}_{n-1}$ of the previous time step. To that end, we wish to employ a standard mark-refine-procedure, however, we have to see to it that the refinement is directed towards $\mathcal{G}_{n-1}$ in the sense of Section 4.2.1. Generally, in order to realize that a mesh $\mathcal{G} \in \mathbb{G}$ is refined towards a mesh $\mathcal{G}' \in \mathbb{G}$, we consider the set of elements in $\mathcal{G}$ of which $\mathcal{G}'$ contains a refinement. More precisely, invoking the notion of the forest associated to a grid (cf. Section 3.1.1 and particularly Definition 3.1.6), we define

$$\mathcal{A}(\mathcal{G}, \mathcal{G}') := \{E \in \mathcal{G} \mid E \in \mathcal{F}(\mathcal{G}') \text{ and } E \text{ has a successor in } \mathcal{F}(\mathcal{G}')\}$$
$$\subset \mathcal{G} \tag{4.33}$$

and directed refinement towards $\mathcal{G}'$ can be achieved by demanding that at least one element $E \in \mathcal{A}(\mathcal{G}, \mathcal{G}')$ is refined in each iterative step. The actual mark-refine-process is then handled by the abstract module MARK_REFINE stated below. In this context, we recall that also the module ST_ADAPTATION involves mesh refinements. However, opposed to the directed refinement towards some target mesh needed for controlling the energy gain, the realization of ST_ADAPTATION presented in Section 4.6.2 also employs a standard ("undirected") refinement for controlling the space error. For convenience, we collect both refinements in the abstract module MARK_REFINE, which optionally receives as an additional argument a target mesh $\mathcal{G}' \in \mathbb{G}$ if directed refinement is desired.

Properties 4.4.12 (MARK_REFINE)
Let $\mathcal{G}$ and $\mathcal{G}'$ be conforming triangulations of Ω. Moreover, let $\{\mathcal{E}(E)\}_{E \in \mathcal{G}} \subset \mathbb{R}_0^+$ be a set of non-negative error indicators where $\mathcal{E}(E)$ is associated to one particular element $E \in \mathcal{G}$. Then, the module

$$\mathcal{G}_* := \text{MARK\_REFINE}(\{\mathcal{E}(E)\}_{E \in \mathcal{G}}, \mathcal{G})$$

generates a set $\mathcal{M}^m \subset \mathcal{G} \times \mathbb{N}$ of marked elements and corresponding element markers. Hereby, the plain set $\mathcal{M} \subset \mathcal{G}$ of marked elements captures the element holding the largest indicator.
If the module is equipped with an additional argument $\mathcal{G}'$, then

$$\mathcal{G}_* := \text{MARK\_REFINE}(\{\mathcal{E}(E)\}_{E \in \mathcal{G}}, \mathcal{G}, \mathcal{G}')$$

generates a set $\mathcal{M}^{\mathrm{m}}$ of marked elements and corresponding element markers and the plain set $\mathcal{M}$ of marked elements contains at least one element of $\mathcal{A}(\mathcal{G}, \mathcal{G}')$ defined in (4.33).

In both cases, the mesh $\mathcal{G}$ is refined via $\mathcal{G}_* := \mathsf{REFINE}(\mathcal{M}^{\mathrm{m}}, \mathcal{G})$. MARK_REFINE terminates and returns the refinement $\mathcal{G}_* \geq \mathcal{G}$.

Remark 4.4.13 (Directed version of MARK_REFINE)

As any target mesh $\mathcal{G}' \in \mathbb{G}$ is associated to a *finite* forest $\mathcal{F}(\mathcal{G}')$ and the directed version MARK_REFINE($\{\mathcal{E}(E)\}_{E \in \mathcal{G}}, \mathcal{G}, \mathcal{G}'$) marks (and thus refines) at least one element in $\mathcal{A}(\mathcal{G}, \mathcal{G}')$ on each call, we see that after finitely many iterations

$$\mathcal{G}_0 := \mathcal{G}, \qquad \mathcal{G}_{k+1} := \mathsf{MARK\_REFINE}(\{\mathcal{E}(E)\}_{E \in \mathcal{G}_k}, \mathcal{G}_k, \mathcal{G}'),$$

a mesh $\mathcal{G}_K \in \mathbb{G}$, which is a refinement of $\mathcal{G}'$, i.e., $\mathcal{G}_K \geq \mathcal{G}'$, is produced. The directed version of MARK_REFINE is used for both establishing the energy gain criterion and reducing the coarsening error if applicable. In the n-th time step, for the latter, it is intuitive to adjust the selection of the marked elements $\mathcal{M}$ to the error indicators $\{\mathcal{E}_c^2(U_{n-1}, \mathcal{G}_n, E)\}_{E \in \mathcal{G}_n}$. However, the condition $\mathcal{M} \cap \mathcal{A}(\mathcal{G}_n, \mathcal{G}_{n-1}) \neq \emptyset$ stated in Properties 4.4.12 must be enforced.

Analogously, we adjust the selection of $\mathcal{M}$ to the local energy gain $\{\mathcal{E}_*^2(U_n, U_{n-1}, \tau_n, \mathcal{G}_n, E)\}_{E \in \mathcal{G}_n}$ to establish the energy gain criterion. We point out that with respect to the energy gain criterion, the critical indicators are *negative* and in order to apply a standard marking strategy, appropriate scaling with -1 should be performed.

The energy gain criterion is ensured at the end of each time step. Rather than shifting this to another abstract module, we employ an explicit **while**-loop (see lines 10–14) in ASTFEM to show openly that the energy gain control involves a call of ST_ADAPTATION.

Remark 4.4.14 (Succession of MARK_COARSEN and CONSISTENCY)

We remark that the modules MARK_COARSEN and CONSISTENCY do not influence each other since whereas MARK_COARSEN only generates a mesh, CONSISTENCY is only concerned with producing a time step size. Thus, the succession of the two modules is arbitrary and we chose to employ MARK_COARSEN first, as it does not provide any error control and thus can be regarded as preliminary in each time step. Opposed to that, CONSISTENCY does provide control for the consistency error and is thus a full-fledged con-

---

Algorithm 5 ASTFEM

1: initialize $\mathcal{G}_{\text{init}}$, $\tau_0 > 0$ and set $t_0 = 0$, $n = 0$

2: $(U_0, \mathcal{G}_0) = \texttt{ADAPT\_INIT}(u_0, \mathcal{G}_{\text{init}})$

3: compute $\tau_*(U_0, f)$

4: **do**

5:    $n = n + 1$

6:    $\mathcal{G}_n = \texttt{MARK\_COARSEN}(U_{n-1}, \mathcal{G}_{n-1})$

7:    $(f_n, \tau_n) = \texttt{CONSISTENCY}(f, t_{n-1}, \tau_{n-1})$

8:    $(U_n, \tau_n, f_n, \mathcal{G}_n) = \texttt{ST\_ADAPTATION}(U_{n-1}, t_{n-1}, \tau_n, f_n,$
         $\mathcal{G}_n, \mathcal{G}_{n-1})$

9:    compute $\{\mathcal{E}_*^2(U_n, U_{n-1}, \tau_n, \mathcal{G}_n, E)\}_{E \in \mathcal{G}_n}$

10:   **while** $\mathcal{E}_*^2(U_n, U_{n-1}, \tau_n, \mathcal{G}_n) < 0$ **do**   ★ control energy
      gain

11:       $\mathcal{G}_n = \texttt{MARKREFINE}(\{\mathcal{E}_*^2(U_n, U_{n-1}, \tau_n, \mathcal{G}_n, E)\}_{E \in \mathcal{G}_n},$
            $\mathcal{G}_n, \mathcal{G}_{n-1})$

12:       $(U_n, \tau_n, f_n, \mathcal{G}_n) = \texttt{ST\_ADAPTATION}(U_{n-1}, t_{n-1}, \tau_n, f_n,$
            $\mathcal{G}_n, \mathcal{G}_{n-1})$

13:       compute $\{\mathcal{E}_*^2(U_n, U_{n-1}, \tau_n, \mathcal{G}_n, E)\}_{E \in \mathcal{G}_n}$

14:   **end while**

15: **while** $t_n = t_{n-1} + \tau_n < T$

---

stituent of each time step. However, from a numerical point of view, switching the two modules is more appropriate since the *spatial* L^2-norm $\|f(s) - f_n\|_{L^2(\Omega)}$ involved in the consistency error is usually evaluated using certain quadrature formulae on the current mesh. Hence, evaluating such spatial norms on a fine mesh *before* coarsening yields more accurate results and should be preferred.

Using the modules $\texttt{ADAPT\_INIT}$, $\texttt{MARK\_COARSEN}$, $\texttt{CONSISTENCY}$, and $\texttt{ST\_ADAPTATION}$ (and implicitly also $\texttt{SOLVE}$ and $\texttt{MARK\_REFINE}$), ASTFEM computes a sequence $\{U_n\}_{n \geq 0}$ of solutions to (EG).

4.5 Termination and convergence into tolerance

In this section, we will exploit the properties of the abstract modules stated in the previous section to show that ASTFEM terminates and that the computed approximation is within any preset tolerance TOL > 0. We follow the lines of [27].

Theorem 4.5.1 (Termination of ASTFEM)
Let $TOL_0, TOL_f, TOL_S > 0$ be positive tolerances and suppose further that the modules ADAPT_INIT, MARK_COARSEN, CONSISTENCY and ST_ADAPTATION possess the properties 4.4.1, 4.4.3, 4.4.6, and 4.4.10 respectively. Then for any initial grid $\mathcal{G}_{init}$ and initial time step size $\tau_0 > 0$, ASTFEM terminates. Particularly, the final time T is reached within a finite number of iterations.

Proof. To prove this theorem, we mainly focus on the outer **do-while**-loop employed in lines 4 through 15 of ASTFEM. In particular, we show that each round of this loop terminates and we utilize properties of the individual time step sizes to ensure global termination of the loop. In the process, we solely exploit the properties of the abstract modules. To allow for this, we inductively verify that the arguments of the individual modules meet the respective prerequisites. To that end, we inductively show for $n - 1 \geq 0$

(a) $\mathcal{G}_{n-1}$ is a conforming triangulation;

(b) $U_{n-1} \in \mathbb{V}_{n-1}$;

(c) $t_{n-1} \in [0, T)$;

(d) $T \geq \tau_{n-1} \geq \min\{\tau_f, \tau_*, \tau_0\}$ or $\tau_{n-1} = T - t_{n-2}$;

where apparently, the alternative $\tau_{n-1} = T - t_{n-2}$ in (d) is void for $n - 1 = 0$.

To start with, we note that in line 1, $\tau_0 > 0$ and $t_0 := 0$ are initialized. Further, from Properties 4.4.1 we observe that ADAPT_INIT($u_0, \mathcal{G}_{init}$) in line 2 terminates producing a conforming triangulation $\mathcal{G}_0$ and $U_0 \in \mathbb{V}_0$. This both establishes above properties for $n = 1$ and shows that the preliminary lines 1 through 3 terminate. We now inductively step from $n - 1$ to n and track the changes performed during a single round of the **do-while**-loop.

By induction hypothesis (a) and (b), we notice that the arguments U_{n-1} and $\mathcal{G}_{n-1}$ are appropriate for the module MARK_COARSEN and hence, by the respective Properties 4.4.3, we conclude that MARK_COARSEN($U_{n-1}, \mathcal{G}_{n-1}$) in line 6 terminates and produces a conforming triangulation $\mathcal{G}_n$; hypothesis (a) is (preliminarily) satisfied for n.

Similarly, the arguments of CONSISTENCY(f, t_{n-1}, τ_{n-1}) in line 7 are suitable by induction hypothesis (c) and (d) and thus, Properties 4.4.6 imply that this module terminates and returns an $f_n \in L^2(\Omega)$. Moreover, it produces a τ_n satisfying

$$\min\{\tau_f, \tau_{n-1}\} \leq \tau_n \leq T - t_{n-1} \qquad \text{or} \qquad \tau_n = T - t_{n-1}$$

and thus, hypothesis (d) is (preliminarily) met for n; $\mathcal{G}_n$ stays untouched.

Next, we observe that U_{n-1} and t_{n-1} are appropriate arguments for ST_ADAPTATION in line 8 by hypothesis (b) and (c), whereas τ_n, f_n and $\mathcal{G}_n$ — which were only created in this very time step — are appropriate due to above considerations. Making use of Properties 4.4.10, we see that ST_ADAPTATION($U_{n-1}, t_{n-1}, \tau_n, f_n, \mathcal{G}_n, \mathcal{G}_{n-1}$) in line 8 terminates. This module may further refine $\mathcal{G}_n$ ((a) stays valid) and returns a $U_n \in \mathbb{V}_n$ ((b) stays valid). Moreover, the time step size τ_n may be reduced. However, in this case we still have $\tau_n \geq \tau_*$ by Properties 4.4.10 and thus also (d) is preserved.

We then turn to the inner **while**-loop in lines 10 through 14 which is supposed to control the energy gain. First we show that any round of this loop terminates. To that end, we note that by Properties 4.4.12, MARK_REFINE in line 11 terminates. It may further refine $\mathcal{G}_n$ conserving (a). As above, ST_ADAPTATION in line 12 terminates and keeps properties (a)-(d) valid. We thus conclude that any single round of the inner **while**-loop terminates preserving induction hypothesis (a)–(d) for n. In order to show that the whole inner **while**-loop of lines 10 through 14 terminates, we show that eventually $\mathcal{E}_*^2(U_n, U_{n-1}, \tau_n, \mathcal{G}_n) \geq 0$. Since ST_ADAPTATION in line 12 may only refine the mesh $\mathcal{G}_n$, we deduce from Properties 4.4.12 that after a finite number of rounds, the directed version of MARK_REFINE in line 11 produces a mesh $\mathcal{G}_n \geq \mathcal{G}_{n-1}$ by, cf. Remark 4.4.13. This gives rise to nested spaces $\mathbb{V}_{n-1} \subset \mathbb{V}_n$ and thus, $\mathcal{E}_*^2(U_n, U_{n-1}, \tau_n, \mathcal{G}_n) \geq 0$ (cf. Section 4.2.1) implies that the inner **while**-loop terminates.

Hence, the whole time step terminates and the newly generated $\mathcal{G}_n$, U_n, τ_n and $t_n = t_{n-1} + \tau_n$ again satisfy above assumptions (a)–(d). We inductively conclude that any time step terminates.

We finally show, that the whole **do-while**-loop (lines 4 through 15) terminates, which is the case if and only if we have $t_N \geq T$ for some $N \in \mathbb{N}$. From hypothesis (d) we recall that all time step sizes particularly satisfy either

$$\tau_n \geq \min\{\tau_f, \tau_*, \tau_0\} \quad \text{or} \quad \tau_n = T - t_{n-1}. \tag{4.34}$$

If for some $N > 0$ the second alternative of (4.34) holds, we have

$$t_N = t_{N-1} + \tau_N = t_{N-1} + T - t_{N-1} = T$$

and consequently, the **do-while**-loop terminates. On the contrary, if all time step sizes meet the first alternative of (4.34), this implies for $N \geq T/\min\{\tau_f, \tau_*, \tau_0\}$

$$t_N = \sum_{n=1}^{N} \tau_n \geq N \min\{\tau_f, \tau_*, \tau_0\} \geq T$$

and also in this case, the **do-while**-loop terminates. $\square$

Again, we emphasize that Theorem 4.5.1 particularly implies that the final time will be reached within a finite number of iterations. This essentially relies on the two minimal time step sizes $\tau_f, \tau_* > 0$ provided by CONSISTENCY and ST_ADAPTATION. It remains to show that Algorithm ASTFEM also provides suitable error control. To that end, we exploit the properties of the abstract modules in combination with error control C introduced in Section 4.3.2.

Theorem 4.5.2 (Convergence of ASTFEM)
The linear interpolation U created from the sequence of solutions $\{U_n\}_{n \geq 0}$ computed by ASTFEM satisfies the error bound

$$\|u - U\|_{\mathbb{W}(0,T)}^2 \leq TOL_0 + TOL_* + \sum_{n=1}^{N} \tau_n (TOL_f + TOL_S).$$

Proof. We show that the individual modules employed in ASTFEM provide the assumptions of error control C. In particular, note that the module MARK_COARSEN does not provide any error control and is hence neglected in the following considerations.

From Properties 4.4.1 we see that ADAPT_INIT produces a grid $\mathcal{G}_0$ and an approximation U_0 of u_0 such that $\mathcal{E}_0^2(u_0, \mathcal{G}_0) \leq$ TOL$_0$ which is assumption (1) of error control C.

We now consider an arbitrary time step n, $n \in \{1, \ldots, N\}$, where $N \in \mathbb{N}$ is the number of time steps needed by ASTFEM for termination. Since in particular any single time step terminates due to Theorem 4.5.1, the condition of the **while**-loop in line 10 implies that the energy gain criterion $\mathcal{E}_*^2(U_n, U_{n-1}, \tau, \mathcal{G}_n) \geq 0$ is satisfied for all $n = 1, \ldots, N$. This is assumption (2) of error control C.

Moreover, we observe that by Properties 4.4.6, the module $(f_n, \tau_n) =$ CONSISTENCY(f, t_{n-1}, τ_{n-1}) in line 7 provides a time-step size τ_n such that $\mathcal{E}_f^2(f, t_{n-1}, \tau) \leq$ TOL$_f$. In the sequel of the n-th time step, this time step size τ_n is not increased and Lemma 4.2.2 therefore implies that $\mathcal{E}_f^2(f, t_{n-1}, \tau_n) \leq$ TOL$_f$ also holds for the time step size τ_n ultimately chosen in the current time step. Hereby, we also established assumption (3) of error control C.

The module executed last of all in the n-th time step is the module ST_ADAPTATION; either in line 8 or in line 12. From Properties 4.4.10 we see that this either ensures

$$\mathcal{E}_{\mathcal{G}}^2(U_n, U_{n-1}, \tau_n, f_n, \mathcal{G}_n) + \mathcal{E}_{c\tau}^2(U_n, U_{n-1}) \leq \text{TOL}_{\text{S}} \qquad \text{if } \tau_n > \tau_*$$

which is assumption (4) of error control C, or

$$\mathcal{E}_{\mathcal{G}}^2(U_n, U_{n-1}, \tau_n, f_n, \mathcal{G}_n) + \mathcal{E}_c^2(U_{n-1}, \mathcal{G}_n) \leq \text{TOL}_{\text{S}} \qquad \text{if } \tau_n \leq \tau_*$$

which is assumption (5) of error control C.

We therefore established all prerequisites of error control C and thus error control C implies the claim. $\qquad\qquad\square$

Choosing the specific positive tolerances TOL$_0$, TOL$_*$, TOL$_f$, and TOL$_{\text{S}}$ appropriately for a given overall tolerance TOL > 0, we see from Theorems 4.5.1 and 4.5.2 that ASTFEM eventually terminates and produces a discrete solution U satisfying the error bound

$$\|u - U\|_{\mathbb{W}(0,T)}^2 \leq \text{TOL}_0 + \text{TOL}_* + \sum_{n=1}^{N} \tau_n(\text{TOL}_f + \text{TOL}_{\text{S}}) \leq \text{TOL}.$$

4.6   Realization of the abstract modules

Finally, in this section we state actual realizations of the abstract modules CONSISTENCY and ST_ADAPTATION defined in Section 4.4, cf. [27]. Those two modules and particularly ST_ADAPTATION are the most important modules of ASTFEM in the sense that they largely determine the behavior of a single time step. For those modules, we show that the presented realization possess the abstract properties. As indicated, the modules ADAPT_INIT, MARK_COARSEN, MARK_REFINE, and SOLVE are considered as black box modules and we do not elaborate on them here.

4.6.1   The module CONSISTENCY

Algorithm 6 states a realization of the module CONSISTENCY, which possesses the corresponding properties from Section 4.4.2 according to Lemma 4.6.1.

Algorithm 6 Module CONSISTENCY (Parameters $\sigma, \delta_1 \in (0,1)$ and $\delta_2 > 1$)

CONSISTENCY(f, t, τ)

1: $\tau = \min\{\tau, T - t\}$

2: compute $\mathcal{E}_f^2(f, t_{n-1}, \tau_n)$

3: **while** $\mathcal{E}_f^2(f, t_{n-1}, \tau_n) < \sigma\, \text{TOL}_f$ and $\tau < T - t$ **do**   ★ enlarge τ

4:      $\tau = \min\{\delta_2\tau, T - t\}$

5:      compute $\mathcal{E}_f^2(f, t_{n-1}, \tau_n)$

6: **end while**

7: **while** $\mathcal{E}_f^2(f, t_{n-1}, \tau_n) > \text{TOL}_f$ **do**   ★ reduce τ

8:      $\tau = \delta_1\tau$

9:      compute $\mathcal{E}_f^2(f, t_{n-1}, \tau_n)$

10: **end while**

11: $\bar{f} = \frac{1}{\tau} \int_t^{t+\tau} f(s)\, ds$

12: **return** $\bar{f}, \tau$

Lemma 4.6.1

For $f \in H^1((0,T), L^2(\Omega))$ we define

$$\tau_f := \frac{\delta_1 \, \textit{TOL}_f}{\mathcal{C}_f \, \|f'\|^2_{L^2((0,T),L^2(\Omega))}}. \tag{4.35}$$

Then the implementation of the module CONSISTENCY *stated in Algorithm 6 satisfies Properties 4.4.6.*

Proof. Let $f \in H^1(0,T; L^2(\Omega))$, $t \in (0,T)$ and $\tau^{\text{in}} \in (0,T)$ as in the prerequisites of Properties 4.4.6. We directly see that the module CONSISTENCY(f,t,τ^{in}) returns the mean value $\bar{f} = \frac{1}{\tau} \int_t^{t+\tau} f(s) \, ds$ defined in line 11 and a time step size τ. Hence, it remains to prove that CONSISTENCY(f,t,τ^{in}) terminates and upon termination,

(a) the time step size τ satisfies either $\min\{\tau_f, \tau^{\text{in}}\} \le \tau \le T - t$ or $\tau = T - t$;

(b) the error control $\mathcal{E}_f^2(f,t,\tau) \le \textit{TOL}_f$ is met.

In line 1, τ is set to either τ^{in} or $T - t$ and (a) is fulfilled in any case. We observe that the module CONSISTENCY consists of two **while**-loops changing the time step size τ. We consider those loops separately.

If the first **while**-loop (lines 3 through 6) is *not* entered, property (a) is obviously valid before entering the second **while**-loop (lines 7 through 10). On the contrary, if the first **while**-loop is entered, the time step size τ is enlarged by a factor $\delta_2 > 1$ in every cycle in line 4, provided it does not exceed $T - t$. This may happen only finitely many times until $\tau \ge T - t$ due to the second condition in line 3. Hence, this first **while**-loop terminates with $\tau \ge \tau^{\text{in}}$ and $\tau \le T - t$ and thus, also in this case condition (a) is valid before entering the second **while**-loop.

If the second **while**-loop is not entered at all, this implies $\mathcal{E}_f^2(f,t,\tau) \le \textit{TOL}_f$ is already satisfied when first hitting line 7. The algorithm thus terminates with (a) and (b) being satisfied.

Before discussing the case of the second **while**-loop being entered, we estimate by Lemma 4.2.2 and using the definition of τ_f in (4.35)

$$\mathcal{E}_f^2(f,t,\tau) \le \mathcal{C}_f \, \tau \, \|f'\|^2_{L^2(0,T;L^2(\Omega))} \le \textit{TOL}_f \qquad \text{for all } \tau \le \tau_f/\delta_1. \tag{4.36}$$

We now turn to the second **while**-loop in lines 7 through 10 and we notice that in each cycle of this loop, the time step size τ is reduced by a fixed factor $\delta_1 < 1$. Consequently, after a finite number of cycles, we have $\tau \leq \tau_f/\delta_1$ and by (4.36) we deduce $\mathcal{E}_f^2(f,t,\tau) \leq \text{TOL}_f$. Hence, the condition in line 7 is violated and the loop terminates. The produced time step size τ is bounded from below by τ_f and thus, also in this case (a) and (b) are satisfied. $\square$

4.6.2 The module ST_ADAPTATION

In Algorithm 7, we state a realization of the module ST_ADAPTATION, which, according to Lemma 4.6.2, satisfies the corresponding properties from Section 4.4.2, see also [27]. ST_ADAPTATION is implemented using an infinite loop (lines 1 through 27) which is exited in line 5 if the space and coarsen-time indicator satisfy the criterion for *big* time steps given in error control C. The second (and last) possibility for leaving the infinite loop is given in line 24. This exit is used if the time step size is *small* in the sense of error control C (i. e., $\tau \leq \tau_*$) and the respective conditions of error control C are satisfied. Moreover, we emphasize that the flow control in ST_ADAPTATION is very adaptive, as we describe in the following: After initially computing the current solution in line 2, the space and coarsen-time indicators are computed (see line 3). If the criterion for big time steps of error control C is not already met, line 6 sees to it that the adaption is carried out with respect to the dominating error source, either space or coarsen-time indicator. In case of a dominating space indicator, the mesh is refined in line 7, whereas in case of a dominating coarsen-time indicator, the associated time and coarsen sub-indicators are computed (see line 9). Then, ST_ADAPTATION distinguishes between big and small time step sizes (see line 10). In case of a big time step size, the adaption is carried out with respect to the dominating time or coarsening indicator (see lines 11 through 16). Opposed to that, in case of a small time step size, the adaption aims at establishing the corresponding criterion of error control C, which involves the space and coarsen indicator. Again, ST_ADAPTATION adapts with respect to the dominating of those two indicators.

In this context, we also refer to Sections 5.2 and 5.3, where ASTFEM is compared to a standard algorithm and above aspects

---

Algorithm 7 Module ST_ADAPTATION

ST_ADAPTATION$(v, t, \tau, \bar{f}, \mathcal{G}, \mathcal{G}^{\mathrm{old}})$

1: **loop forever**
2:     $U = \mathrm{SOLVE}(v, \bar{f}, \tau, \mathcal{G})$
3:     compute $\{\mathcal{E}^2_{\mathcal{G}}(U, v, \tau, \bar{f}, \mathcal{G}, E)\}_{E \in \mathcal{G}}$ and $\mathcal{E}^2_{c\tau}(U, v)$
4:     **if** $\mathcal{E}^2_{\mathcal{G}}(U, v, \tau, \bar{f}, \mathcal{G}) + \mathcal{E}^2_{c\tau}(U, v) \leq \mathrm{TOL_S}$ **then**
5:        **break**   ★ std. exit
6:     **else if** $\mathcal{E}^2_{\mathcal{G}}(U, v, \tau, \bar{f}, \mathcal{G}) > \mathcal{E}^2_{c\tau}(U, v)$ **then**   ★ $\mathcal{E}_{\mathcal{G}}$ dominates
7:        $\mathcal{G} = \mathrm{MARK\_REFINE}(\{\mathcal{E}^2_{\mathcal{G}}(U, v, \tau, \bar{f}, \mathcal{G}, E)\}_{E \in \mathcal{G}}, \mathcal{G})$ $\boxed{\text{A}}$
8:     **else**   ★ $\mathcal{E}_{c\tau}$ dominates
9:        compute $\mathcal{E}^2_{\tau}(U, v, \mathcal{G})$ and $\{\mathcal{E}^2_c(v, \mathcal{G}, E)\}_{E \in \mathcal{G}}$
10:        **if** $\tau > \tau_*$ **then**
11:           **if** $\mathcal{E}^2_{\tau}(U, v, \mathcal{G}) > \mathcal{E}^2_c(v, \mathcal{G})$ **then**   ★ $\mathcal{E}_{\tau}$ dominates
12:              $\tau = \max\{\delta_1 \tau, \tau_*\}$ $\boxed{\text{B}}$
13:              $\bar{f} = \frac{1}{\tau} \int_t^{t+\tau} f(s)\, ds$
14:           **else**   ★ $\mathcal{E}_c$ dominates
15:              $\mathcal{G} = \mathrm{MARK\_REFINE}(\{\mathcal{E}^2_c(v, \mathcal{G}, E)\}_{E \in \mathcal{G}}, \mathcal{G}, \mathcal{G}^{\mathrm{old}})$ $\boxed{\text{C}}$
16:           **end if**
17:        **else if** $\mathcal{E}^2_{\mathcal{G}}(U, v, \tau, \bar{f}, \mathcal{G}) + \mathcal{E}^2_c(v, \mathcal{G}) > \mathrm{TOL_S}$ **then**
18:           **if** $\mathcal{E}^2_c(v, \mathcal{G}) > \mathcal{E}^2_{\mathcal{G}}(U, v, \tau, \bar{f}, \mathcal{G})$ **then**   ★ $\mathcal{E}_c$ dominates
19:              $\mathcal{G} = \mathrm{MARK\_REFINE}(\{\mathcal{E}^2_c(v, \mathcal{G}, E)\}_{E \in \mathcal{G}}, \mathcal{G}, \mathcal{G}^{\mathrm{old}})$ $\boxed{\text{D}}$
20:           **else**   ★ $\mathcal{E}_{\mathcal{G}}$ dominates
21:              $\mathcal{G} = \mathrm{MARK\_REFINE}(\{\mathcal{E}^2_{\mathcal{G}}(U, v, \tau, \bar{f}, \mathcal{G}, E)\}_{E \in \mathcal{G}}, \mathcal{G})$ $\boxed{\text{E}}$
22:           **end if**
23:        **else**
24:           **break**   ★ non-std. exit
25:        **end if**
26:     **end if**
27: **end loop forever**
28: **return** $U, \tau, \bar{f}, \mathcal{G}$

---

are particularly emphasized.

Lemma 4.6.2 (Termination of ST_ADAPTATION)
The module ST_ADAPTATION stated in Algorithm 7 satisfies Properties 4.4.10.

Proof. Let $\mathcal{G}^{\text{in}}$ and $\mathcal{G}^{\text{old}}$ be conforming triangulations of Ω. Further, let $v^{\text{in}} \in \mathbb{V}(\mathcal{G}^{\text{old}})$, $\bar{f}^{\text{in}} \in L^2(\Omega)$ and $t, \tau^{\text{in}} \in (0, T)$ such that $t + \tau^{\text{in}} \in (0, T)$. We show that the module

$$(U, \tau, \bar{f}, \mathcal{G}) = \text{ST\_ADAPTATION}(v^{\text{in}}, t, \tau^{\text{in}}, \bar{f}^{\text{in}}, \mathcal{G}^{\text{in}}, \mathcal{G}^{\text{old}})$$

meets the properties stated in Properties 4.4.10.

Solution. We first observe that the **loop forever**-loop in Algorithm 7 is an infinite loop which may only be left through one of the two break statements in lines 5 and 24. Assuming that the algorithm terminates, this implies that one of those two statements is carried out directly before returning U, τ, $\bar{f}$ and $\mathcal{G}$. Moreover, we note that modifications of τ or $\mathcal{G}$ are exclusively performed in lines 7, 12, 15, 19, and 21. The flow control inside the **loop forever**-loop consists of several **if**-statements and ensures that none of those lines is executed after SOLVE was called in line 2 of the *final* cycle. Hence, the returned U indeed is the solution corresponding to the returned τ and $\mathcal{G}$.

Termination. We next show that Algorithm 7 terminates. To that end, we argue by contradiction and assume that Algorithm 7 does not terminate, which is the case if and only if the statements in line 5 and 24 are never executed. The flow within **loop forever** is governed by several **if**-conditions, which lead to the execution of exactly one of the statements marked by A,...,E in every cycle. Since the employed modules SOLVE and MARK_REFINE are deterministic by Properties 4.4.9 and 4.4.12, the assumption that **loop forever** does not terminate implies that at least one of the statements A,...,E is executed infinitely many times. In the following, we deduce a contradiction from each of the assumptions that A,...,E is executed infinitely many times.

$\boxed{\text{B}}$ is executed infinitely many times: Since the time step size τ is reduced by a factor δ_1 or directly set to τ_* in each execution of B, we have $\tau \leq \tau_*$ after a finite number of calls. Thus, due to the **if**-switch in line 10, statement B is not called anymore; a contradiction. Since the time step size τ is exclusively modified in statement B, this particularly guarantees that the time step size will eventually be fixed.

$\boxed{\text{C}}$ is executed infinitely many times: The **if**-switch in line 6 ensures $\mathcal{E}_{\mathcal{G}}^2(U, v, \tau, \bar{f}, \mathcal{G}) \leq \mathcal{E}_{c\tau}^2(U, v)$ whenever C is reached and thus, we have

$$\mathcal{E}_{\mathcal{G}}^2(U, v, \tau, \bar{f}, \mathcal{G}) + \mathcal{E}_{c\tau}^2(U, v) \leq 2\,\mathcal{E}_{c\tau}^2(U, v).$$

Splitting $\mathcal{E}_{c\tau}^2(U, v)$ into $\mathcal{E}_{\tau}^2(U, v, \mathcal{G})$ and $\mathcal{E}_c^2(v, \mathcal{G})$, we further estimate

$$\leq 2\big(\mathcal{E}_{\tau}^2(U, v, \mathcal{G}) + \mathcal{E}_c^2(v, \mathcal{G})\big)$$

and since the **else**-switch in line 14 ensures $\mathcal{E}_{\tau}^2(U, v, \mathcal{G}) \leq \mathcal{E}_c^2(v, \mathcal{G})$, we have

$$\leq 4\,\mathcal{E}_c^2(v, \mathcal{G}). \tag{4.37}$$

The directed version of MARK_REFINE in C produces a grid $\mathcal{G} \geq \mathcal{G}^{\text{old}}$ after finitely many calls by Properties 4.4.12. This implies $\mathcal{E}_c^2(v, \mathcal{G}) = 0$ and consequently, estimate (4.37) implies

$$\mathcal{E}_{\mathcal{G}}^2(U, v, \tau, \bar{f}, \mathcal{G}) + \mathcal{E}_{c\tau}^2(U, v) \leq 0 < \text{TOL}_{\mathsf{S}}$$

and the **while**-loop exits through the break in line 5; a contradiction.

$\boxed{\text{D}}$ is executed infinitely many times: Again, we exploit that the directed version of MARK_REFINE produces a grid $\mathcal{G} \geq \mathcal{G}^{\text{old}}$ within finitely many iterations implying $\mathcal{E}_c^2(v, \mathcal{G}) = 0$. Thanks to the **if**-switch in line 18 we have

$$\mathcal{E}_{\mathcal{G}}^2(U, v, \tau, \bar{f}, \mathcal{G}) + \mathcal{E}_c^2(v, \mathcal{G}) < 2\mathcal{E}_c^2(v, \mathcal{G}) = 0 < \text{TOL}_{\mathsf{S}}$$

after finitely many iterations, which contradicts the **if**-switch in line 17.

$\boxed{\text{E}}$ is executed infinitely many times: The time step size τ is exclusively modified in line 12 marked by B, which we recall is only executed finitely many times. Hence, the time step size will eventually be fixed. This implies that the elliptic problem of the current time step does not change anymore, cf. Remark 3.3.22.

Observing that also C and D are only called finitely many times, ST_ADAPTATION eventually becomes a standard algorithm for solving the current elliptic problem and coincides with Algorithm 1. As the employed modules and the error estimator $\mathcal{E}_\mathcal{G}$ meet the assumptions of Theorem 3.3.21, we may employ the convergence for elliptic problems from Theorem 3.3.21 to deduce $\mathcal{E}_\mathcal{G}^2(U, v, \tau, \bar{f}, \mathcal{G}) \to 0$. Thanks to the **else**-switch in line 20 we have

$$\mathcal{E}_\mathcal{G}^2(U, v, \tau, \bar{f}, \mathcal{G}) + \mathcal{E}_c^2(v, \mathcal{G}) \leq 2\mathcal{E}_\mathcal{G}^2(U, v, \tau, \bar{f}, \mathcal{G}) < \mathtt{TOL_S}$$

after a finite number of iterations which again contradicts the **if**-switch in line 17.

$\boxed{\text{A}}$ is executed infinitely many times: As in above contradiction for E, the time step size will eventually be fixed and the convergence for elliptic problems from Theorem 3.3.21 implies $\mathcal{E}_\mathcal{G}^2(U, v, \tau, \bar{f}, \mathcal{G}) \to 0$. Hence, from the **if**-switch in line 6 we have

$$\mathcal{E}_\mathcal{G}^2(U, v, \tau, \bar{f}, \mathcal{G}) + \mathcal{E}_{c\tau}^2(U, v) < 2\,\mathcal{E}_\mathcal{G}^2(U, v, \tau, \bar{f}, \mathcal{G}) < \mathtt{TOL_S}$$

after a finite number of iterations which contradicts the **if**-switch in line 4.

We deduce that ST_ADAPTATION terminates.

Error control. If the algorithm terminates exiting the **loop forever**-loop via the **break** in line 5, the **if**-switch in line 4 ensures that

$$\mathcal{E}_\mathcal{G}^2(U, v, \tau, \bar{f}, \mathcal{G}) + \mathcal{E}_{c\tau}^2(U, v) \leq \mathtt{TOL_S},$$

which is the desired bound in case $\tau > \tau_*$. As we cannot exclude the case where the algorithm exits through this standard exit for $\tau \leq \tau_*$, we employ the estimate $\mathcal{E}_c^2(U, v) \leq \mathcal{E}_{c\tau}^2(U, v)$ to also establish the desired bound for this case

$$\mathcal{E}_\mathcal{G}^2(U, v, \tau, \bar{f}, \mathcal{G}) + \mathcal{E}_c^2(U, v) \leq \mathtt{TOL_S}.$$

Opposed to that, if the algorithm exits via the non-standard exit in line 24, the **if**-switch in line 10 implies that $\tau \leq \tau_*$. Moreover, the **if**-switch in line 17 ensures that

$$\mathcal{E}_\mathcal{G}^2(U, v, \tau, \bar{f}, \mathcal{G}) + \mathcal{E}_c^2(v, \mathcal{G}) \leq \mathtt{TOL_S}.$$

Time step size. During ST_ADAPTATION, the time-step size is only modified in line 12 (marked by B) where it is reduced only such that $\tau \geq \tau_*$. Hence it holds $\tau_* \leq \tau \leq \tau^{\text{in}}$. If $\tau^{\text{in}} < \tau_*$, the modification in line 12 is never executed due to the **if**-switch in line 10 and hence, the time step size is not changed at all. $\qquad \square$

Chapter 5

Numerical aspects of ASTFEM

After having focused on the rather theoretical aspects of ASTFEM — particularly termination and convergence into tolerance — in the last chapter, we now switch our point of view to more practical aspects. In Section 5.1 we consider various coarsening strategies and finally introduce the very flexible "multi-coarsening", which we also employ in the implementations underlying the computational results of Sections 5.3 and 6.5. In Section 5.2 we briefly review a standard algorithm for solving parabolic problems and compare it to ASTFEM. Finally, in Section 5.3 we present numerical results which both show that ASTFEM performs well and also explicitly point out advantages over the standard algorithm.

5.1 Coarsening in ASTFEM

In this section, we present the coarsening strategy employed in the implementation of ASTFEM which was used for the computations presented in Section 5.3.

The general goal of coarsening in adaptive algorithms for time dependent problems is to remove unnecessary degrees of freedom, particularly ones that were needed at earlier times but are not needed *anymore*. This feature is very important as the spatial structure of the solution generally changes with the passing of time. Such changes in general require local refinements of the underlying mesh in order to satisfy given tolerances, on the other hand they also render unnecessary previously needed degrees of freedom. The textbook examples for such phenomena are moving fronts or moving peaks, cf. Section 5.3.

In general, we split the coarsening procedure into a part MARK selecting a subset $\mathcal{M} \subset \mathcal{G}$ of marked elements with associated element markers $m(E) \in \mathbb{N}$ for each $E \in \mathcal{M}$, and a part COARSEN implementing the actual coarsening. For the latter, we employ the black box module introduced in Section 3.1.1. We recall that this module is equipped with a triangulation $\mathcal{G}$ and a set $\mathcal{M}^m \subset \mathcal{G} \times \mathbb{N}$ of marked elements $E \in \mathcal{G}$ and corresponding element markers $m(E) \in \mathbb{N}$. For convenience, we use the notation $\mathcal{M}^k \subset \mathcal{G} \times \mathbb{N}$ for marked elements with all element markers being equal to k, i.e., $\mathcal{M}^k := \{(E, k) \mid E \in \mathcal{M} \subset \mathcal{G}\}$. Given a conforming triangulation $\mathcal{G}$ and a set $\mathcal{M}^m \subset \mathcal{G} \times \mathbb{N}$ of marked elements $E \in \mathcal{M} \subset \mathcal{G}$ and corresponding element markers $m(E) \in \mathbb{N}$, $\mathcal{G}_- = \mathsf{COARSEN}(\mathcal{M}^m, \mathcal{G})$ outputs a conforming triangulation $\mathcal{G}_- \leq \mathcal{G}$ where any element $E \in \mathcal{M}$ is coarsened at most $m(E)$ times. Recall further that since unmarked elements $E \in \mathcal{G} \setminus \mathcal{M}$ must not be coarsened, in general not all marked elements $E \in \mathcal{M}$ can actually be coarsened $m(E)$ times due to preserving conformity, cf. Section 3.1.1.

From this splitting employing the black box module COARSEN, we realize that we influence and determine the whole coarsening process by defining the module MARK. In particular, we define the set $\mathcal{M} \subset \mathcal{G}$ of marked elements and the associated element markers $m(E)$, $E \in \mathcal{M}$, generated by MARK.

In the following, we present different strategies for the implementation of this module. As before, we first focus on strategies which associate with each marked element $E \in \mathcal{M}$ the same element marker $m(E) = k$, particularly we consider $k = 1$. In this case, we conveniently use the notation $\mathcal{M}^k$ for the set of marked elements and corresponding markers. However, the coarsening strategy employed in the implementation of ASTFEM used for generating the computational results of Section 5.3, is more advanced and involves individual element markers $m(E)$ for each element.

Recalling from Section 4.4.2 that $\mathsf{MARK\_COARSEN}(U, \mathcal{G})$ does not provide any error control, we are at liberty to choose any strategy for mesh coarsening.

5.1.1 Uniform coarsening

The most basic strategies for coarsening a given mesh $\mathcal{G}$ involve *global* coarsening. As we implement a coarsening strategy by defining the set $\mathcal{M}^m$ of marked elements and corresponding element markers generated by the module MARK, we obtain global coarsening by marking the entire mesh, i.e., we set $\mathcal{M} := \mathcal{G}$. Then,

$$\mathcal{G}_- := \text{COARSEN}(\mathcal{M}^1, \mathcal{G})$$

produces the coarsened mesh $\mathcal{G}_-$ with $\mathcal{G}_{\text{init}} \leq \mathcal{G}_- \leq \mathcal{G}$. However, we note that even in a global coarsening step, in general not all elements $E \in \mathcal{G}$ are coarsened since conformity has to be preserved, compare Figure 5.1. A natural generalization of employing one global coarsening step is to globally coarsen a mesh k times for $k \geq 2$, i.e., we set $\mathcal{G}_- := \text{COARSEN}(\mathcal{M}^k, \mathcal{G})$.

Yet another possibility for a uniform coarsening strategy is to coarsen a given mesh $\mathcal{G}$ directly into the underlying macro triangulation $\mathcal{G}_{\text{init}}$. In this case, the coarsening of $\mathcal{G}$ is actually independent of $\mathcal{G}$ and the whole module MARK_COARSEN$(U, \mathcal{G})$ can be defined directly via

$$\mathcal{G}_{\text{init}} =: \text{MARK\_COARSEN}(U, \mathcal{G})$$

without splitting the process into MARK and COARSEN subroutines.

Remark 5.1.1
Note that the presented strategies do not coarsen a given mesh $\mathcal{G}$ uniformly in the sense that each $E \in \mathcal{G}$ is coarsened equally many times, see also Figure 5.1. The term "uniform" in this context rather refers to the fact, that MARK treats all elements $E \in \mathcal{G}$ equally, i.e., no decisions are involved whether or not a given element should be marked for coarsening.

5.1.2 Adaptive coarsening

In contrast to the presented uniform coarsening strategies, we now consider strategies which selectively coarsen a mesh $\mathcal{G}$ into some

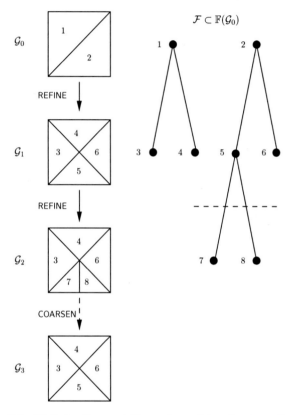

Figure 5.1: Global refinement (from $\mathcal{G}_0$ to $\mathcal{G}_1$), selective refinement (from $\mathcal{G}_1$ to $\mathcal{G}_2$), and global coarsening (from $\mathcal{G}_2$ to $\mathcal{G}_3$) as well as corresponding forest $\mathcal{F}$. Note that the global coarsening of $\mathcal{G}_2$ does not coarsen elements 3, 4 and 6 due to preserving conformity. The coarsening is indicated in $\mathcal{F}$ as dashed line "$- -$".

$\mathcal{G}_- \leq \mathcal{G}$. The idea behind this is that the coarsened mesh $\mathcal{G}_-$ shall be chosen such that a given function $V \in \mathbb{V}(\mathcal{G})$ can be "approximated well" in the corresponding coarsened finite element space $\mathbb{V}(\mathcal{G}_-) \subset \mathbb{V}(\mathcal{G})$. More precisely, given a function $V \in \mathbb{V}(\mathcal{G})$, a new grid $\mathcal{G}_- \leq \mathcal{G}$ is generated such that $\|V - V_-\|$ is small. Hereby, $\|\cdot\|$ is an arbitrary norm on $\mathbb{V}(\mathcal{G})$ and $V_- \in \mathbb{V}(\mathcal{G}_-)$ is a "representation" of V in the coarsened space, e. g., V_- can be chosen as $V_- := \mathcal{P}V$ for some interpolation or projection operator $\mathcal{P} : \mathbb{V}(\mathcal{G}) \to \mathbb{V}(\mathcal{G}_-)$. We emphasize that such a coarsening is goal-oriented for *one particular* function $V \in \mathbb{V}(\mathcal{G})$. The coarsening thus depends on both the mesh $\mathcal{G}$ as well as the particular function $V \in \mathbb{V}(\mathcal{G})$. Additionally, it hinges on the choice of the norm and the representation V_-.

To implement such a coarsening strategy, we again focus on the module MARK selecting the set $\mathcal{M}^m$ of marked elements and associated element markers which will be used as input for COARSEN. From the foregoing considerations, we equip MARK with both a mesh $\mathcal{G}$ and a finite element function $V \in \mathbb{V}(\mathcal{G})$ as arguments. In order to get a representation of a function $V \in \mathbb{V}(\mathcal{G})$ in the finite element space corresponding to the coarsened mesh $\mathcal{G}_- \leq \mathcal{G}$, we employ the Lagrange interpolation operator $\mathcal{I}_{\mathbb{V}(\mathcal{G}_-)} : \mathbb{V}(\mathcal{G}) \to \mathbb{V}(\mathcal{G}_-)$ introduced in Theorem 3.1.16. As a norm to measure the mismatch between a $V \in \mathbb{V}(\mathcal{G})$ and its representation $V_- \in \mathbb{V}(\mathcal{G}_-)$, we choose the energy norm $\|\cdot\|_\Omega$. Then

$$\mathcal{M}^m := \mathsf{MARK}(V, \mathcal{G})$$

aims at selecting elements and corresponding element markers such that the coarsening error for V satisfies

$$\left\| V - \mathcal{I}_{\mathbb{V}(\mathcal{G}_-)} V \right\|_\Omega^2 \leq \mathtt{TOL_c} \tag{5.1}$$

for the associated mesh

$$\mathcal{G}_- = \mathsf{COARSEN}(\mathcal{M}^m, \mathcal{G})$$

and a non-negative coarsening tolerance $\mathtt{TOL_c} \geq 0$. Moreover, $\mathcal{M}^m$ should be chosen such that the number $\#\mathcal{G} - \#\mathcal{G}_-$ of removed elements is preferably large. We explicitly point out that $\mathtt{TOL_c}$ is only used as a parameter for the coarsening procedure. In particular, it is *not* related to any error control for $\|u - U\|_{\mathbb{W}(0,T)}$ in algorithm ASTFEM.

Remark 5.1.2 (Admissible elements)
We recall that in ASTFEM, coarsening is used to generate an initial guess for the current mesh $\mathcal{G}_n$ by coarsening the previous mesh $\mathcal{G}_{n-1}$. To avoid coarsening/refinement loops and acknowledging the fact that the solution does not change a lot in two consecutive time steps (provided the time step size is chosen appropriately), it seems appropriate to forbid coarsening of elements that were only just refined. To be more precise, we consider the beginning of the n-th time step where $\mathcal{G}_n$ is generated from $\mathcal{G}_{n-1}$ using information from U_{n-1} via

$$\mathcal{G}_n = \texttt{MARK\_COARSEN}(U_{n-1}, \mathcal{G}_{n-1}).$$

In this process, elements $E \in \mathcal{G}_{n-1}$ that were newly created in $\mathcal{G}_{n-1}$ are called *not admissible for coarsening*. Hence, in the n-th time step, the set of admissible elements is given by

$$\mathcal{A}_n := \mathcal{G}_{n-1} \setminus \{E \in \mathcal{G}_{n-1} \mid E \text{ is not admissible}\}.$$

Moreover, it might be appropriate to additionally consider the element error indicator
$\mathcal{E}_{\mathcal{G}}^2(U_{n-1}, U_{n-2}, \tau_{n-1}, f_{n-1}, \mathcal{G}_{n-1}, E)$ — which indicates the local space error on E in time step $n-1$ — in order to decide whether an element $E \in \mathcal{G}_{n-1}$ shall be allowed for coarsening: Assuming that the solution does not change much from time step $n-1$ to time step n, coarsening an element with relatively large error indicator will probably be reverted in the course of time step n by refining the produced element again. With this in mind, we may further restrict the set of admissible elements and define

$$\mathcal{A}_n^* := \mathcal{A}_n \setminus \{E \in \mathcal{G}_{n-1} \mid \mathcal{E}_{\mathcal{G}}^2(U_{n-1}, U_{n-2}, \tau_{n-1}, f_{n-1}, \mathcal{G}_{n-1}, E)$$
$$\text{is too large}\}$$

as admissible elements. A popular way to specify if the error indicator is "too large" is to check whether

$$\mathcal{E}_{\mathcal{G}}^2(U_{n-1}, U_{n-2}, \tau_{n-1}, f_{n-1}, \mathcal{G}_{n-1}, E)$$
$$> \beta \max_{E' \in \mathcal{G}_{n-1}} \mathcal{E}_{\mathcal{G}}^2(U_{n-1}, U_{n-2}, \tau_{n-1}, f_{n-1}, \mathcal{G}_{n-1}, E')$$

for some parameter $\beta \in (0, 1)$. In the actual implementation of the coarsening routine used for the simulations presented in Sections 5.3 and 6.5, we used $\mathcal{A}_n$ as admissible elements. For ease of presentation, we do not pursue the concept of admissible elements explicitly in what follows. However, we are at liberty to implicitly restrict the set $\mathcal{M}$ of marked elements to $\mathcal{A}_n$ or $\mathcal{A}_n^*$ in each time step n.

In the following paragraphs on single coarsening and multi-coarsening, we introduce strategies for constructing sets $\mathcal{M}^m$ of marked elements and associated element markers such that (5.1) will be satisfied. We will often use a curly minus "$\sim$" as subscript to elements or meshes. We chose this symbol to consistently indicate that the bearer of "$\sim$" is an auxiliary object in the course of constructing the ultimate coarsened mesh $\mathcal{G}_-$. The objects with subscript "$\sim$" are usually associated to coarser meshes.

Single coarsening

We first consider coarsening strategies employing the same marker $m(E) = k$ for all $E \in \mathcal{M}$, particularly, we consider $k = 1$. Hence, we elaborate on how the plain set $\mathcal{M} \subset \mathcal{G}$ of marked elements — which directly gives rise to $\mathcal{M}^k$ — can be generated such that the tolerance criterion (5.1) is met for the associated coarsened mesh

$$\mathcal{G}_- = \mathsf{COARSEN}(\mathcal{M}^k, \mathcal{G}).$$

For this purpose, let $\mathcal{G}_\sim \in \mathbb{G}$ be the global coarsening of $\mathcal{G}$, i. e.,

$$\mathcal{G}_\sim := \mathsf{COARSEN}(\mathcal{G}^1, \mathcal{G})$$

and assume we have at hand the local coarsening error terms $\left\| \left| V - \mathcal{I}_{\mathbb{V}(\mathcal{G}_\sim)} V \right| \right\|_E^2$ for all $E \in \mathcal{G}$. We understand $\mathcal{G}_\sim$ as an auxiliary coarsening and we are looking for a coarsened mesh $\mathcal{G}_-$ with $\mathcal{G}_\sim \leq \mathcal{G}_- \leq \mathcal{G}$ satisfying (5.1).

If the tolerance criterion (5.1) is readily satisfied for the global coarsening $\mathcal{G}_\sim$, we mark all elements for coarsening, i. e., $\mathcal{M} = \mathcal{G}$, and employ $\mathcal{G}_\sim =: \mathcal{G}_-$ as coarsened mesh.

In the opposed case $\left\| \left| V - \mathcal{I}_{\mathbb{V}(\mathcal{G}_\sim)} V \right| \right\|_\Omega^2 > \mathsf{TOL}_c$ however, we have to select a (proper) subset $\mathcal{M} \subset \mathcal{G}$ giving rise to $\mathcal{G}_-$ via

$$\mathcal{G}_- = \mathsf{COARSEN}(\mathcal{M}^1, \mathcal{G})$$

instead of employing the global coarsening. To construct the set $\mathcal{M}$ of marked elements, we use the information about the local coarsening errors and we select $\mathcal{M}$ such that

$$\sum_{E \in \mathcal{M}} \left\| \left\| V - \mathcal{I}_{\mathbb{V}(\mathcal{G}_\sim)} V \right\| \right\|_E^2 \leq \mathtt{TOL_c}. \tag{5.2}$$

Based on that, the actual coarsening $\mathcal{G}_- = \mathsf{COARSEN}(\mathcal{M}^1, \mathcal{G})$ is generated, implying $\mathcal{G}_\sim \leq \mathcal{G}_- \leq \mathcal{G}$. Again, we emphasize that in general, not all marked elements are actually coarsened. Since the Lagrange interpolation operator is a local operator, we have for the interpolation to the coarse mesh $\mathcal{G}_-$

$$\left\| \left\| V - \mathcal{I}_{\mathbb{V}(\mathcal{G}_-)} V \right\| \right\|_E^2 = \begin{cases} 0 & \begin{array}{l} \text{for all elements } E \in \mathcal{G} \\ \text{that are } not \text{ coarsened,} \\ \text{i.\,e., } E \in \mathcal{G} \text{ and } E \in \mathcal{G}_-; \end{array} \\ \left\| \left\| V - \mathcal{I}_{\mathbb{V}(\mathcal{G}_\sim)} V \right\| \right\|_E^2 & \begin{array}{l} \text{for all coarsened ele-} \\ \text{ments, i.\,e., } E \in \mathcal{G} \\ \text{but } E \notin \mathcal{G}_-. \end{array} \end{cases}$$

Using this property and keeping in mind that only marked elements may be coarsened, we conclude by (5.2) that also in this case, the tolerance criterion (5.1) is satisfied:

$$\left\| \left\| V - \mathcal{I}_{\mathbb{V}(\mathcal{G}_-)} V \right\| \right\|_\Omega^2 \leq \sum_{E \in \mathcal{M}} \left\| \left\| V - \mathcal{I}_{\mathbb{V}(\mathcal{G}_\sim)} V \right\| \right\|_E^2 \leq \mathtt{TOL_c}.$$

In this light, we aim at constructing a set of marked elements $\mathcal{M}$ such that criterion (5.2) is satisfied. In order to generate a preferably large set $\mathcal{M}$, we construct $\mathcal{M}$ iteratively using an "inverse Dörfler" strategy explained in the following. We choose a parameter $\gamma > 0$ and in a first round, only mark elements E with

$$\left\| \left\| V - \mathcal{I}_{\mathbb{V}(\mathcal{G}_\sim)} V \right\| \right\|_E^2 \leq \gamma \frac{\mathtt{TOL_c}}{\#\mathcal{G}}. \tag{5.3}$$

Note that during one round, elements satisfying (5.3) are only added to $\mathcal{M}$ *preliminary*. At the end of each round, it is checked if (5.2) holds for the set $\mathcal{M}$, which was preliminary enlarged during the foregoing round. If (5.2) is satisfied, the newly marked elements are permanently accepted in $\mathcal{M}$ and a new round starts with an increased parameter γ. Opposed to that, if (5.2) is violated, all elements marked during the previous round are removed from $\mathcal{M}$ and the iteration ends. We emphasize, that only criterion (5.2) provides error control. The criterion stated in (5.3) is purely heuristic and is exclusively employed to suggest elements for being marked.

Hence, we generated a set $\mathcal{M}$ of marked elements satisfying (5.2) as desired.

Remark 5.1.3 (Choice of criterion (5.3))
Even though at first sight, the criterion stated in (5.3) reminds of the equidistribution strategy (compare the marking strategies described in Section 3.3.5 on page 78), the scaling with $1/\#\mathcal{G}$ on the right hand side of (5.3) is only used to allow for using the same criterion and same threshold γ on various meshes. On a very fine grid, we expect $\left\| V - \mathcal{I}_{V(\mathcal{G}_\sim)} V \right\|_E$ to be rather small as $|E|$ is small and since in this case, $\#\mathcal{G}$ is quite large, the scaling seems appropriate.

Remark 5.1.4 (Inverse Dörfler marking)
Comparing the presented "inverse Dörfler" marking and the Dörfler marking briefly described in Section 3.3.5 on page 79, we note that those strategies are quite different. However, they do have in common the fact that they both use greedy algorithms to generate a set $\mathcal{M}$ of marked elements. Thereby, the Dörfler strategy aims at producing a preferably *small* set $\mathcal{M}$, whereas the *inverse* Dörfler strategy pursues the generation of a preferably *large* set $\mathcal{M}$. As Dörfler's strategy is the only widespread marking strategy employing a greedy approach, we chose this naming for the introduced strategy.

The foregoing presentation is *starting point oriented* in the sense that we consider the effect of *choosing an element* $E \in \mathcal{G}$ and coarsening it. Hence, the focus lies on the elements of the original mesh $\mathcal{G}$, which is the starting point in the coarsening process. Opposed to that, we may change our point of view and consider the effect of *employing a particular* $E_\sim \in \mathcal{G}_\sim$ *as new element*, i.e., coarsening all $E \in \mathcal{G}$ with $E \subset E_\sim$ into $E_\sim \in \mathcal{G}_\sim$. This approach is *goal-oriented* in the sense that the focus lies on the (potential) coarsening $\mathcal{G}_\sim$, which is the objective of the coarsening process.

The difference of these two points of view is rather subtle, but nevertheless, it is very convenient to employ the goal-oriented view in the considerations regarding multi-coarsening. To point out the difference more explicitly, we define the set of subelements in the

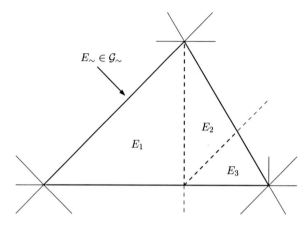

Figure 5.2: Fine grid $\mathcal{G}$ (dashed) and coarse grid $\mathcal{G}_\sim$ (solid). For the coarse grid element $E_\sim \in \mathcal{G}_\sim$, the set $E_{\mathcal{G}}^{\mathrm{sub}}(E_\sim)$ consists of the fine grid elements $E_1, E_2, E_3 \in \mathcal{G}$ contained in $E_\sim$, i.e., $E_{\mathcal{G}}^{\mathrm{sub}}(E_\sim) = \{E_1, E_2, E_3\}$.

fine mesh $\mathcal{G}$ of $E_\sim \in \mathcal{G}_\sim$ as

$$E_{\mathcal{G}}^{\mathrm{sub}}(E_\sim) := \{E \in \mathcal{G} \mid E \subset E_\sim\},$$

an illustration of this set is given in Figure 5.2. Let $E_\sim \in \mathcal{G}_\sim$ be a given element in the coarse mesh. Then, the starting-point-oriented approach focuses on the coarsening error for each individual $E \in E_{\mathcal{G}}^{\mathrm{sub}}(E_\sim)$ separately and considers for each of those elements whether or not it should be coarsened. On the other hand, the goal-oriented approach focuses on the *complete* error introduced by coarsening the whole *set* of subelements $E_{\mathcal{G}}^{\mathrm{sub}}(E_\sim)$ into $E_\sim \in \mathcal{G}_\sim$.

Multi-coarsening

The introduced single coarsening procedure is only able to coarsen an element *once*, even if it produces no coarsening error at all. To overcome this drawback, one might employ a coarser $\mathcal{G}_\sim$ (explicitly $\mathcal{G}_\sim := \mathsf{COARSEN}(\mathcal{G}^k, \mathcal{G})$) and a set $\mathcal{M}^k$ with $k \geq 2$ in the foregoing considerations. However, the single coarsening strategy only evaluates local aspects of a suggested coarsening to a *fixed* $\mathcal{G}_\sim$. Opposed to that, the presented multi-coarsening does not employ a *fixed* suggestion $\mathcal{G}_\sim$ but rather considers *all possible* coarsenings of $\mathcal{G}$. As it is inconvenient to store information about all those potential coarsening errors on the elements of the un-coarsened mesh, we employ a goal-oriented approach and rather focus on the coarsened element $E_\sim$ than on the original elements $E \in E_\mathcal{G}^{\mathrm{sub}}(E_\sim)$.

For this purpose, we consider the the forest $\mathcal{F}(\mathcal{G}) \in \mathbb{F}(\mathcal{G}_{\mathrm{init}})$ associated to the triangulation $\mathcal{G}$ (cf. Definition 3.1.6), as it contains all possible coarsenings of $\mathcal{G}$. For any element $E_\sim \in \mathcal{F}(\mathcal{G})$, we consider the coarsening error introduced by employing $E_\sim$ as a new leaf element, i.e., coarsening all $E \in E_\mathcal{G}^{\mathrm{sub}}(E_\sim)$ into $E_\sim$. To be more precise on the produced coarsening error, let $E_\sim \in \mathcal{F}(\mathcal{G})$ and let $\mathcal{G}(E_\sim) \subset \mathcal{F}(\mathcal{G})$ be a coarsening of $\mathcal{G}$ which contains $E_\sim$ as an element, i.e., $E_\sim \in \mathcal{G}(E_\sim)$. Conveniently, we abbreviate $\mathcal{I}_{E_\sim} := \mathcal{I}_{\mathbb{V}(\mathcal{G}(E_\sim))}$ for the Lagrange interpolation operator mapping to a finite element space whose underlying mesh contains $E_\sim$. Since we are only interested in the *local* coarsening error $\|V - \mathcal{I}_{E_\sim} V\|_{E_\sim}$, the precise (global) choice of $\mathcal{G}(E_\sim)$ — which also determines $\mathbb{V}(\mathcal{G}(E_\sim))$ — is not important as the Lagrange interpolation operator is local. Having at hand $\|V - \mathcal{I}_{E_\sim} V\|_{E_\sim}$ for all $E_\sim \in \mathcal{F}(\mathcal{G})$, we aim at selecting a set $\mathcal{T} \subset \mathcal{F}(\mathcal{G})$ of *target elements*, that shall be employed in the desired coarsening $\mathcal{G}_-$ of $\mathcal{G}$. This is to be understood in the sense that for any $E_\sim \in \mathcal{T}$, all elements in $E_\mathcal{G}^{\mathrm{sub}}(E_\sim)$ shall be coarsened to $E_\sim$.

Employing $\mathcal{T}$. Before presenting a strategy how to construct the set of target elements $\mathcal{T}$, we elaborate on how to employ $\mathcal{T}$ for coarsening $\mathcal{G}$ into a $\mathcal{G}_-$. For this purpose, let $\mathcal{T} \subset \mathcal{F}(\mathcal{G})$. We then wish to coarsen $\mathcal{G}$ such that all $E \in E_\mathcal{G}^{\mathrm{sub}}(E_\sim)$ are coarsened into $E_\sim$ for all $E_\sim \in \mathcal{T}$. However, we cannot employ the black box module COARSEN directly, since it expects a set $\mathcal{M}^{\mathrm{m}} \subset \mathcal{G} \times \mathbb{N}$ of marked elements and corresponding element markers, whereas $\mathcal{T} \subset \mathcal{F}(\mathcal{G})$ consists of target elements. Nevertheless, we may easily produce a set $\mathcal{M}^{\mathrm{m}}$ of elements $E \in \mathcal{G}$ and associated element markers $m(E)$ from $\mathcal{T}$ by employing the notion of the generation g of an element respectively node in the forest $\mathcal{F}(\mathcal{G})$, (cf. Definitions 3.1.6 and 3.1.5). We note that in order to produce E from $E_\sim$, $g(E) - g(E_\sim)$ refinements are needed. Vice versa, we realize that also $g(E) - g(E_\sim)$ coarsenings are needed to coarsen an element E into $E_\sim$. With this in mind, we transform a target element $E_\sim \in \mathcal{T}$ into the desired pairs of elements and associated element markers by setting

$$\mathcal{M}_{E_\sim} := \Big\{ \big(E, m(E)\big) \in \mathcal{G} \times \mathbb{N} \mid E \in E_\mathcal{G}^{\mathrm{sub}}(E_\sim) \text{ and}$$

$$m(E) := g(E) - g(E_\sim) \Big\}.$$

Note that if all elements in $\mathcal{M}_{E_\sim}$ are actually coarsened $m(E)$ times, this indeed produces $E_\sim$. Hence, $\mathcal{M}_{E_\sim}$ is suited to transfer the desired information to the black box module COARSEN. From the sets $\mathcal{M}_{E_\sim}$ associated to one target element $E_\sim$, we finally define

$$\mathcal{M}^{\mathrm{m}} = \mathcal{M}^{\mathrm{m}}(\mathcal{T}) := \bigcup_{E_\sim \in \mathcal{T}} \mathcal{M}_{E_\sim}$$

and set

$$\mathcal{G}_- := \mathsf{COARSEN}(\mathcal{M}^{\mathrm{m}}, \mathcal{G}).$$

Constructing $\mathcal{T}$. In the following, we present a strategy for selecting the target elements $E_\sim \in \mathcal{T}$. Before proceeding with this, we recall that the aim of coarsening is to produce a mesh $\mathcal{G}_-$ by removing as many elements as possible while satisfying the demanded tolerance (5.1). As selecting a target element $E_\sim$ aims at removing all subelements $E_\mathcal{G}^{\mathrm{sub}}(E_\sim) \subset \mathcal{G}$, it seems appropriate to put into perspective the number of subelements and the induced coarsening error. With this information, MARK can decide whether or not an element $E_\sim \in \mathcal{F}(\mathcal{G})$ shall be included in $\mathcal{T}$ by considering the ratio

$$\frac{\| V - \mathcal{I}_{E_\sim} V \|_{E_\sim}^2}{\# E_\mathcal{G}^{\mathrm{sub}}(E_\sim)}$$

of induced coarsening error and number of subelements of $E_\sim$. To construct the set $\mathcal{T} \subset \mathcal{F}(\mathcal{G})$, we employ an inverse Dörfler marking to select target elements in a set $\mathcal{T} \subset \mathcal{F}(\mathcal{G})$ with respect to the error-to-subelements-ratios, compare the paragraph on single coarsening on page 175ff. In the following, we describe a single round of the inverse Dörfler marking employing a threshold parameter $\gamma > 0$. Analogously to the inverse Dörfler marking in the context of single coarsening on page 176, a single round only adds target elements preliminary to $\mathcal{T}$. It is only after the whole round is finished, that those target elements are either permanently accepted in $\mathcal{T}$ or removed again.

Algorithm 8 Algorithms used for recursive traverse in multi-coarsening. REC_TRAVERSE is launched on a macro element $E \in \mathcal{G}_{\text{init}}$.

INNER(E)
 if no predecessor of E is contained in $\mathcal{T}$ **then**
  if (5.4) is satisfied **then**
   preliminary add E to $\mathcal{T}$
  end if
 else   predecessor E_P of E in $\mathcal{T}$
  if $\|V - \mathcal{I}_E V\|_E > \|V - \mathcal{I}_{E_P} V\|_{E_P}$ **then**
   remove predecessor E_P from $\mathcal{T}$ again
  end if
 end if

REC_TRAVERSE(E)
 if E has successor **then**
  INNER(E)
  REC_TRAVERSE(left_successor(E))
  REC_TRAVERSE(right_successor(E))
 end if

In each single round, we traverse the forest $\mathcal{F}(\mathcal{G})$ by recursively traversing each tree $\mathbb{F}(E_\sim)$ rooted in an element $E_\sim \in \mathcal{G}_{\text{init}}$ of the macro triangulation. The recursive routine, which is also stated in Algorithm 8, is triggered on the root of the current tree. For any element $E_\sim \in \mathcal{F}(\mathcal{G})$, it checks whether a predecessor of $E_\sim$ is already contained in $\mathcal{T}$. If so, $E_\sim$ itself may not be added. Otherwise, we check whether

$$\frac{\|V - \mathcal{I}_{E_\sim} V\|_{E_\sim}^2}{\#E_\mathcal{G}^{\text{sub}}(E_\sim)} \leq \gamma \frac{\text{TOL}_c}{\#\mathcal{G}} \tag{5.4}$$

is satisfied, compare also criterion (5.3) for single coarsening and Remark 5.1.3. If criterion (5.4) is not satisfied, the recursive routine is called on the left successor of $E_\sim$ before being called on the right successor of $E_\sim$. If $E_\sim$ is a leaf and hence does not possess any successor, the recursive routine ends. In the more interesting case when criterion (5.4) is satisfied, $E_\sim$ is (preliminary) added to $\mathcal{T}$. In this case, we do not allow that a successor of $E_\sim$ is added to $\mathcal{T}$. More interestingly however, a target element $E_\sim$ preliminary added to $\mathcal{T}$ may directly be removed from $\mathcal{T}$ again before the current round ends. This happens if any successor $E'_\sim$ of $E_\sim$ produces a larger coarsening error than $E_\sim$ itself, i. e.,

$$\left\|\left|V - \mathcal{I}_{E'_\sim} V\right\|\right|_{E'_\sim} > \left\|V - \mathcal{I}_{E_\sim} V\right\|_{E_\sim}.$$

This proceeding is substantiated in the following observation. For a target element $E_\sim \in \mathcal{T}$, we desire that all subelements $E \in E_{\mathcal{G}}^{\mathrm{sub}}(E_\sim)$ are coarsened $m(E) = g(E) - g(E_\sim)$ times. However, due to the enforced preserving of conformity, we cannot guarantee that an element $E \in E_{\mathcal{G}}^{\mathrm{sub}}(E_\sim)$ is actually coarsened $m(E)$ times and in general, E may be coarsened only m' times with $0 < m' < m(E)$. Hence, in order to guarantee that the ultimate coarsening $\mathcal{G}_- \leq \mathcal{G}$ satisfies (5.1), we need to provide control also for $\left\|\left|V - \mathcal{I}_{E'_\sim} V\right\|\right|_{E'_\sim}$, where $E'_\sim$ is an intermediate element between E and $E_\sim$, i. e., $E'_\sim$ is any successor of a selected target element $E_\sim \in \mathcal{T}$. Note that even though $E'_\sim$ is from a finer mesh than $E_\sim$, in general, we cannot expect $\left\|\left|V - \mathcal{I}_{E'_\sim} V\right\|\right|_{E'_\sim} \leq \left\|V - \mathcal{I}_{E_\sim} V\right\|_{E_\sim}$, compare Figure 5.3.

Having completed one of these rounds, we check if

$$\sum_{E_\sim \in \mathcal{T}} \left\|V - \mathcal{I}_{E_\sim} V\right\|_{E_\sim}^2 \leq \mathrm{TOL_c} \tag{5.5}$$

is satisfied for the preliminary enlarged $\mathcal{T}$. If so, the target elements temporarily added to $\mathcal{T}$ in the foregoing round are added permanently, otherwise, they are all removed again and the iteration ends. Again, we emphasize that error control is exclusively provided by (5.5), whereas criterion (5.4) is solely employed for suggesting potential target elements.

As described before, we employ $\mathcal{T}$ to derive the set $\mathcal{M}^{\mathrm{m}}$ of marked elements and associated element markers and we generate the coarsened mesh $\mathcal{G}_- := \mathsf{COARSEN}(\mathcal{M}^{\mathrm{m}}, \mathcal{G})$. As $\left\|\left|V - \mathcal{I}_{E'_\sim} V\right\|\right|_{E'_\sim}$

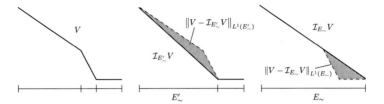

Figure 5.3: 1d example of finite element interpolation to different grids: original function V on fine grid (left), Lagrange interpolation to intermediate grid (middle) and coarse grid (right). In the middle and right figure, the original function is indicated as dashed line and the interpolation error to the intermediate element $E'_\sim$ (middle) respectively coarse element $E_\sim$ (right) is shaded. From basic geometric considerations we conclude that the interpolation error $\left\|V - \mathcal{I}_{E'_\sim} V\right\|_{L^1(E'_\sim)}$ for the intermediate grid is greater than the error $\left\|V - \mathcal{I}_{E_\sim} V\right\|_{L^1(E_\sim)}$ for the coarse grid.

$\leq \left\|V - \mathcal{I}_{E_\sim} V\right\|_{E_\sim}$ for all successors $E'_\sim$ of $E_\sim \in \mathcal{T}$, the largest coarsening error is produced if all marked elements are actually coarsened into the respective target element $E_\sim \in \mathcal{T}$. Hence, we finally obtain by (5.5)

$$\left\|\left\|V - \mathcal{I}_{V(\mathcal{G}_-)} V\right\|\right\|_\Omega^2 \leq \sum_{E_\sim \in \mathcal{T}} \left\|V - \mathcal{I}_{E_\sim} V\right\|_{E_\sim}^2 \leq \mathtt{TOL_c}.$$

Remark 5.1.5
It is reasonable to choose the Lagrange interpolation operator to define the "representation" of a fine grid function on a coarse grid, since in finite element codes, this operator can be implemented very efficiently. For details, we refer to [57, Sections 3.3 and 3.5].

5.2   Comparison with classical algorithm

In this section, we compare ASTFEM with a standard adaptive algorithm for parabolic problems, which is stated in Algorithm 9. This algorithm, to which we refer as CLASSIC, is similar to the one presented by Chen and Feng in [14]. Unless stated otherwise, the line numbers in the following always refer to CLASSIC.

Algorithm 9 CLASSIC (Parameters $\sigma, \delta_1 \in (0,1)$ and $\delta_2 > 1$)

1: initialize $\mathcal{G}_{\text{init}}$, $\tau_0 > 0$ and set $t_0 = 0$, $n = 0$

2: $(U_0, \mathcal{G}_0) = \texttt{ADAPT\_INIT}(u_0, \mathcal{G}_{\text{init}})$

3: **do**

4:     $n = n + 1$

5:     set $\mathcal{G}_n := \mathcal{G}_{n-1}$ and $\tau_n := \tau_{n-1}$

6:     $U_n = \texttt{SOLVE}(U_{n-1}, f_n, \tau_n, \mathcal{G}_n)$

7:     compute $\mathcal{E}_f^2(f, t_{n-1}, \tau_n)$ and $\mathcal{E}_\tau^2(U_n, U_{n-1}, \mathcal{G}_n)$

8:     **while** $\mathcal{E}_f^2(f, t_{n-1}, \tau_n) > \texttt{TOL}_f$ or $\mathcal{E}_\tau^2(U_n, U_{n-1}, \mathcal{G}_n) > \texttt{TOL}_t$ **do**

9:         $\tau_n = \delta_1 \tau_n$

10:        $U_n = \texttt{SOLVE}(U_{n-1}, f_n, \tau_n, \mathcal{G}_n)$

11:        compute $\mathcal{E}_f^2(f, t_{n-1}, \tau_n)$ and $\mathcal{E}_\tau^2(U_n, U_{n-1}, \mathcal{G}_n)$

12:     **end while**

13:     compute $\{\mathcal{E}_\mathcal{G}^2(U_n, U_{n-1}, \tau_n, f_n, \mathcal{G}_n, E)\}_{E \in \mathcal{G}}$

14:     **while** $\mathcal{E}_\mathcal{G}^2(U_n, U_{n-1}, \tau_n, f_n, \mathcal{G}_n) > \texttt{TOL}_\mathcal{G}$ **do**

15:        $\mathcal{G}_n = \texttt{MARK\_REFINE}(\{\mathcal{E}_\mathcal{G}^2(U_n, U_{n-1}, \tau_n, f_n, \mathcal{G}_n, E)\}_{E \in \mathcal{G}_n}, \mathcal{G}_n)$

16:        $U_n = \texttt{SOLVE}(U_{n-1}, f_n, \tau_n, \mathcal{G}_n)$

17:        compute $\mathcal{E}_\tau^2(U_n, U_{n-1}, \mathcal{G}_n)$

18:        **while** $\mathcal{E}_\tau^2(U_n, U_{n-1}, \mathcal{G}_n) > \texttt{TOL}_t$ **do**

19:           $\tau_n = \delta_1 \tau_n$

20:           $U_n = \texttt{SOLVE}(U_{n-1}, f_n, \tau_n, \mathcal{G}_n)$

21:           compute $\mathcal{E}_\tau^2(U_n, U_{n-1}, \mathcal{G}_n)$

22:        **end while**

23:        compute $\{\mathcal{E}_\mathcal{G}^2(U_n, U_{n-1}, \tau_n, f_n, \mathcal{G}_n, E)\}_{E \in \mathcal{G}}$

24:     **end while**

25:     $\mathcal{G}_n = \texttt{MARK\_COARSEN}(U_n, \mathcal{G}_n)$

26:     $U_n = \texttt{SOLVE}(U_{n-1}, f_n, \tau_n, \mathcal{G}_n)$

27:     **if** $\mathcal{E}_f^2(f, t_{n-1}, \tau_n) < \sigma \texttt{TOL}_f$ and $\mathcal{E}_\tau^2(U_n, U_{n-1}, \mathcal{G}_n) < \sigma \texttt{TOL}_t$ **then**

28:        $\tau_n := \delta_2 \tau_n$

29:     **end if**

30: **while** $t_n = t_{n-1} + \tau_n < T$

Before we proceed with the comparison of CLASSIC and AST-FEM, we briefly examine the structure of CLASSIC. First, the initial mesh $\mathcal{G}_{\text{init}}$ and the initial time step size τ_0 are initialized and the time t_0 and the time step counter n are set. Then, the module ADAPT_INIT is employed to generate a mesh $\mathcal{G}_0$ and an appropriate approximation $U_0 \in \mathbb{V}(\mathcal{G}_0)$ of the initial data u_0.

After these preliminaries, the actual time stepping begins. We consider a single time step from $n-1$ to n. As an initial guess for the mesh $\mathcal{G}_n$ in the n-th time step, CLASSIC employs the mesh of the previous time step, likewise the previous time step size is reused, see line 5. Then, CLASSIC immediately solves for the solution U_n in the current time step (line 6). In lines 8 through 12, CLASSIC reduces the time step size τ_n until consistency and time tolerance are met. In the **while**-loop starting in line 14, the space tolerance is checked and, if necessary, the mesh is refined and the corresponding solution is computed, see lines 15 and 16. Since changing the mesh might increase the time error (cf. Section 4.2.6), CLASSIC then again reduces the time step size until the temporal tolerance is reached again. This is done in the inner **while**-loop in lines 18 through 22. Having completed the loop controlling the spatial error (lines 14 through 24), the mesh is coarsened in line 25 and the solution U_n is computed for the coarsened mesh $\mathcal{G}_n$. As a last action in a time step, the time step size is increased, provided both consistency and time errors distinctly deceed their respective tolerances, see lines 27 through 29.

We now emphasize the most significant differences of algorithms CLASSIC and ASTFEM.

1. **Choice of initial guess for τ_n.** ASTFEM employs the module CONSISTENCY to provide an initial guess for τ_n that is adapted to the current behavior of the right hand side f in the sense that the consistency error satisfies its tolerance while not deceeding it distinctly. CLASSIC on the other hand adjusts the time step size such that both the consistency and the time error meet their tolerances. Since particularly the time error is considered, solving for U_n is required in each cycle of this loop (lines 8 through 12). Note also that increasing the time step size is only possible (once) in the end of each time step.

2. **Choice of initial guess for $\mathcal{G}_n$.** We note that whereas AST-FEM coarsens the ultimate mesh of the previous time step, $\mathcal{G}_{n-1}$, in the beginning of the current time step to produce a first guess for $\mathcal{G}_n$, CLASSIC basically does the same by coarsening $\mathcal{G}_{n-1}$ in the end of time step $n-1$ and employing this coarsened mesh as an initial guess for $\mathcal{G}_n$. However, there is a large philosophical gap between the two concepts: ASTFEM closes the time step $n-1$ with a non-coarsened mesh $\mathcal{G}_{n-1}$ and a corresponding solution U_{n-1} in the "non-coarsened space" $\mathbb{V}(\mathcal{G}_{n-1})$. Opposed to that, CLASSIC carries out the coarsening *before* closing time step $n-1$ and hence, the final mesh $\mathcal{G}_{n-1}$ is already coarsened and most importantly, also the final solution U_{n-1} of time step $n-1$ corresponds to the "coarsened space" $\mathbb{V}(\mathcal{G}_{n-1})$. In this light, we see that this strategy discards previously achieved precision of the solution in the un-coarsened space by coarsening the space and re-computing the solution on the "poorer" (coarsened) space. This directly implies, that the bound on the space error indicator, which is established in each time step n, is generally *not* valid for the ultimate solution U_n of a given time step. Since also the time error indicator is coupled to the mesh (cf. Section 4.2.6), the same applies here. Thus, the coarsening step introduces a new error source and it is necessary that the coarsening routine does provide error control. More precisely, let $\tilde{U}_n \in \mathbb{V}(\tilde{\mathcal{G}}_n)$ denote the intermediate solution of the n-th time step directly *before* the coarsening step. Then, the coarsened mesh $\mathcal{G}_n \leq \tilde{\mathcal{G}}_n$ has to be chosen such that

$$\frac{1}{\tau_n} \left\| \tilde{U}_n - \mathcal{I}_{\mathbb{V}(\mathcal{G}_n)} \tilde{U}_n \right\|^2 + \left\| \tilde{U}_n - \mathcal{I}_{\mathbb{V}(\mathcal{G}_n)} \tilde{U}_n \right\|_\Omega^2 \leq \texttt{TOL}_c$$

for some coarsening tolerance $\texttt{TOL}_c \geq 0$. Particularly, this coarsening tolerance enters the estimate of the discretization error. For details, we refer to [14]. On the other hand, we emphasize that this coarsening strategy implies nested finite element spaces $\mathbb{V}(\mathcal{G}_{n-1}) \subset \mathbb{V}(\mathcal{G}_n)$ throughout the time step until the coarsening procedure is reached. Particularly, this implies that $\|U_n - U_{n-1}\|_\Omega$ is a plain time error and hence, a splitting into coarsening and time sub-indicators as employed in ASTFEM is not necessary, cf. Section 4.2.5.

Besides, we point out yet another direct consequence of the different coarsening strategies. For this purpose, let $\mathcal{G}_{n-1}$ be coarsened into some initial guess $\mathcal{G}_{n,0}$ for the mesh of the n-th time step. We assume that $\mathcal{G}_{n,0}$ is refined in the course of the n-th time step into some $\mathcal{G}_n \geq \mathcal{G}_{n-1}$, which particularly is a *refinement* of $\mathcal{G}_{n-1}$. This is possible in both algorithms CLASSIC and ASTFEM. However, whereas AST-FEM is able to resolve the full (un-coarsened) information of $U_{n-1} \in \mathbb{V}(\mathcal{G}_{n-1}) \subset \mathbb{V}(\mathcal{G}_n)$, the coarsening procedure in CLASSIC does not allow for this kind of information recovery and the n-th time step may only resolve the coarsened solution U_{n-1}.

3. **Termination.** We recall that the termination of ASTFEM and especially the fact that the desired final time T is reached, was established by employing a minimal time step size $\tau_* > 0$ granted by special error control, cf. Sections 4.3 and 4.5. As CLASSIC does not employ a minimal time step size, it is unknown whether CLASSIC terminates, particularly it is unknown whether the final time T will be reached. However, Chen and Feng showed in [14], that an algorithm similar to CLASSIC finishes each time step in a finite number of iterations.

4. **Flow control and adaptive decisions.** As in numerical experiments CLASSIC however is observed to reach the final time, the most significant difference between CLASSIC and ASTFEM in this respect lies in the flow controls, which are both constructed such that certain tolerances will be reached eventually. With this in mind, we particularly elaborate on the various tolerance criteria enforced in CLASSIC and AST-FEM. Regarding the consistency error, we notice that both algorithms explicitly demand

$$\mathcal{E}_f^2(f, t_{n-1}, \tau_n) \leq \text{TOL}_f$$

in each time step. More interestingly, we observe that while CLASSIC calls for

$$\mathcal{E}_{\mathcal{G}}^2(U_n, U_{n-1}, \tau_n, f_n, \mathcal{G}_n) \leq \text{TOL}_{\mathcal{G}},$$
$$\mathcal{E}_\tau^2(U_n, U_{n-1}, \mathcal{G}_n) \leq \text{TOL}_t,$$

compare line 14 respectively lines 8 and 18 in CLASSIC, ASTFEM only demands the *cumulated* tolerance

$$\mathcal{E}_{\mathcal{G}}^2(U_n, U_{n-1}, \tau_n, f_n, \mathcal{G}_n) + \mathcal{E}_{c\tau}^2(U_n, U_{n-1}) \leq \text{TOL}_S$$

respectively, in case of non-standard time steps,

$$\mathcal{E}_{\mathcal{G}}^2(U_n, U_{n-1}, \tau_n, f_n, \mathcal{G}_n) + \mathcal{E}_c^2(U_n, U_{n-1}) \leq \text{TOL}_S,$$

compare line 4 respectively line 17 of ST_ADAPTATION. This enables ASTFEM to counterbalance low coarsen-time error indicators by allowing for additional space error. On the other hand, CLASSIC must meet the explicit tolerance for the space error estimator even if the time error distinctly deceeds its granted tolerance. Moreover, we observe that CLASSIC always carries out adaptions in the same chronology. Firstly, the time step size is adjusted such that the consistency and time tolerances are satisfied. Then, provided the space error does not yet fulfill its tolerance criterion, the mesh is refined before the time error is checked again. In contrast, ASTFEM firstly considers a cumulated error and if necessary, it splits the cumulated error into its constituents and takes appropriate action to reduce the *dominating* part, cf. Section 4.6.2. In view of this, adaptive decisions not only cover *if* a certain action is triggered, but also *which* action is most appropriate.

In the following section, we present a numerical example, which particularly emphasizes the differences in flow control and adaptive decisions.

5.3   Numerical results

The foundation of the numerical examples presented in the following, is an implementation of ASTFEM in the C programming language. For this implementation, we strongly employed the finite element toolbox ALBERTA, which is described in detail in [57]. Apart from the official ALBERTA release, we also used a multi-mesh extension, which was generously provided by ALBERTA coauthor A. Schmidt, cf. [55, 54]. The multi-mesh tools allow for simultaneously dealing with finite element functions associated to different meshes. This is particularly useful, as the Euler-Galerkin discretization (EG) involves the solutions $U_n \in \mathbb{V}(\mathcal{G}_n)$ and $U_{n-1} \in \mathbb{V}(\mathcal{G}_{n-1})$ with different meshes $\mathcal{G}_n$ and $\mathcal{G}_{n-1}$. Using the multi-mesh tools, we can cope directly with the term $\langle U_n - U_{n-1}, V \rangle$ for $V \in \mathbb{V}_n$ and do not have to employ the L^2-projection $\Pi_n U_{n-1} \in \mathbb{V}_n$ of U_{n-1} into $\mathbb{V}_n$ to compute $\langle U_n - U_{n-1}, V \rangle$ via

$$\langle U_n - U_{n-1}, V \rangle = \langle U_n - \Pi_n U_{n-1}, V \rangle \qquad \text{for all } V \in \mathbb{V}_n,$$

or even restrict ourselves to nested spaces $\mathbb{V}_{n-1} \subset \mathbb{V}_n$ by employing a coarsening strategy as in CLASSIC. As a coarsening strategy, we employed the multi-coarsening presented in Section 5.1.2.

Wandering peak

We now state the employed data of the first example: For the elliptic differential operator in (2.1), we choose $\mathcal{L} := -\Delta$, i.e., $\mathbf{A} = \mathrm{id}$, $\mathbf{b} = 0$, and $c = 0$. Hence, equation (2.1) becomes the heat equation, which we consider on the time interval $(0, 1)$ and the two-dimensional spatial domain $\Omega := (-1, 1) \times (-1, 1)$. The right hand side f is chosen such that the exact solution is given by

$$u(x, t) = \alpha(t)\, e^{-\beta\left((x_1 - t + 0.5)^2 + (x_2 - t + 0.5)^2\right)} \qquad \text{with}$$
$$\alpha(t) = 1 - e^{-\gamma(t - 0.5)^2}$$

and parameters $\beta = 25$ and $\gamma = 10^4$. Considering the heat equation particularly implies that the bilinear form $\mathcal{B}$ is symmetric and hence, the minimal time step size is given as in (4.22) by

$$\tau_* = \left(10\left(\|U_0\|_\Omega^2 + \|f\|_{L^2(0,1;L^2(\Omega))}^2\right)\right)^{-1} \mathrm{TOL}_*,$$

where $\|U_0\|_\Omega^2 = 3.14$ and $\|f\|_{L^2(0,1;L^2(\Omega))} = 3.18 \cdot 10^2$.

The behavior of the function u can be described as a spatial peak moving at constant speed across the domain Ω. The peak experiences a rapid exponential drop around time $t = 0.5$ where it flattens to $u(x, 0.5) \equiv 0$. After this drop, the peak recovers as fast as it dropped and continues its uniform movement. This example, which is also employed in [14], is chosen such that various aspects of adaptivity can be examined, as we will do in the following.

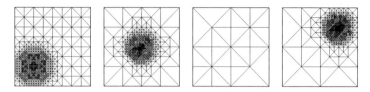

Figure 5.4: Adapted meshes generated by ASTFEM for $t =$ 0, 0.48512, 0.50062, 0.99912 respectively.

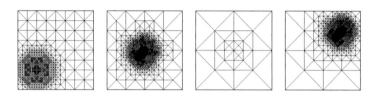

Figure 5.5: Adapted meshes generated by CLASSIC for $t =$ 0, 0.48512, 0.50062, 0.99962 respectively.

We employ algorithm ASTFEM to compute an approximation to u and examine whether ASTFEM satisfies the expectations regarding adaptivity stated below. Before going into this, we introduce the required parameters for ASTFEM. We choose an initial time step size $\tau_0 = 10^{-3}$. As parameters for the time adaptation, we set $\delta_1 := 0.5$ (factor for decreasing the time step size), $\delta_2 := 2$ (factor for increasing the time step size) and $\sigma := 0.5$ (threshold factor for deciding if the time step size is increased). The employed tolerances are $\text{TOL}_0 := 2.5 \cdot 10^{-6}$, $\text{TOL}_f := 2.5 \cdot 10^{-3}$ and $\text{TOL}_S := 5 \cdot 10^{-3}$. Moreover, we set $\text{TOL}_* := 2.5 \cdot 10^{-3}$ implying a minimal time step size $\tau_* = 7.78 \cdot 10^{-7}$. Recall that τ_* is only valid with respect to the time discretization error $\mathcal{E}_\tau$ in the sense that the module ST_ADAPTATION — which is the one module dealing with $\mathcal{E}_\tau$ — does not reduce τ below τ_*. However, the consistency error may require that the module CONSISTENCY does produce time step sizes deceeding τ_*.

As a marking strategy we used the maximum strategy with $\theta = 0.95$ for the generation of the initial mesh and $\theta = 0.8$ for the MARK_REFINE modules.

In the following, we state various expectations an adaptive algorithm should meet for the given example. Conveniently, we directly check whether the results produced by ASTFEM answer those expectations.

• Spacial adaptivity should capture the shape of the peak and locally refine the mesh. — Indeed, Figure 5.4 clearly shows that the mesh is refined locally capturing the shape of the peak.

• The coarsening strategy should take advantage of the peak's movement and coarsen the mesh in previously refined areas. — In line with expectations, Figure 5.4 also shows that the mesh is coarsened in previously refined areas after the peak moved on.

• The coarsening strategy should also detect the drop to a constant function around $t = 0.5$ and coarsen the grid accordingly. — The heavy coarsening around $t = 0.5$ can be seen in the second and third mesh of Figure 5.4 depicting the meshes corresponding to times $t = 0.48512$ respectively $t = 0.50062$. According to expectation, the mesh corresponding to $t = 0.50062$ is indeed very coarse.

• Temporal adaptivity should detect the uniform behavior in time away from $t = 0.5$ and produce a constant time step size. — Figure 5.6 clearly shows that a constant time step size is employed away from $t = 0.5$.

• Temporal adaptivity should further resolve the rapid drop around $t = 0.5$ by sufficiently reducing the time step size. — According to expectations, we see from Figure 5.6 that the time step size is heavily reduced around $t = 0.5$.

• Once the solution moves uniformly again, time adaptivity should re-establish the previously used time step size. — Also this aspect of time adaptivity is verified in Figure 5.6.

We conclude, that all expectations are answered by ASTFEM indicating that the employed methods are indeed reasonable. Moreover, we particularly point out that the smallest time step size actually employed in ASTFEM is $6.25 \cdot 10^{-5}$ and hence much bigger than τ_*. Consequently, ASTFEM never uses the special treatment

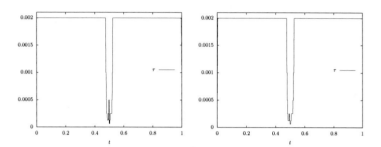

Figure 5.6: Time step sizes used by ASTFEM (left) and CLASSIC (right).

of the τ_*-case and τ_* does not influence the algorithmic flow of ASTFEM in this example but only serves as a theoretical tool for guaranteeing termination within tolerance.

Next, we wish to compare the two algorithms ASTFEM and CLASSIC. For this cause, we solve the same problem employing algorithm CLASSIC with the same values for the initial and consistency tolerances as used in ASTFEM. Since CLASSIC requires individual tolerances for the space and time error estimators, we split TOL_S equally into $TOL_{\mathcal{G}} := TOL_t := 2.5 \cdot 10^{-3}$. Moreover, we allow for an additional coarsening tolerance of $TOL_c := 3 \cdot 10^{-5}$.

Figure 5.5 shows the adapted meshes produced by CLASSIC at (nearly) the same times as the meshes generated by ASTFEM (see Figure 5.4). We observe that CLASSIC also captures well the movement of the peak and its flattening around $t = 0.5$ and produces meshes which are quite similar to the meshes generated by ASTFEM. However, we notice that the meshes of CLASSIC are finer, see also the number of degrees of freedom (DOFs) in Figure 5.10.

Figure 5.7 shows that the consistency error behaves almost identically for both algorithms. Away from $t = 0.5$, the consistency error is constant with $\sigma TOL_f < \mathcal{E}_f^2 < TOL_f$. However, we point out that the consistency error may well drop below σTOL_f even though CONSISTENCY tries to enlarge the time step size in this situation. This is because the enlarged time step size possibly induces a consistency error exceeding TOL_f and is hence reduced again, compare

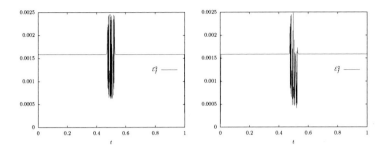

Figure 5.7: Consistency error in ASTFEM (left) and CLASSIC (right).

the realization of CONSISTENCY in Algorithm 6.

In Figure 5.8 displaying the combined coarsen time error respectively the time error, we see that the time error distinctly deceeds its granted tolerance. Particularly, in CLASSIC it never exceeds $3.5 \cdot 10^{-4}$ while the respective tolerance allows up to $\mathcal{E}_\tau^2 = 2.5 \cdot 10^{-3}$. This happens since the consistency error does not allow for a larger time step size and hence, the time step size is exclusively controlled by the consistency error in this example. As algorithm CLASSIC does not provide any means for alternatively using the unneeded shares of tolerance, a considerable part of $\mathtt{TOL}_f$ is wasted. Moreover, comparing the combined coarsen-time error of ASTFEM and the time error of CLASSIC in Figure 5.8, we observe that those quantities are of similar order. This suggests that the coarsening part of the coarsen-time error indicator is quite small.

Most importantly however, we see in Figure 5.9 that ASTFEM adapts according to the cumulated error $\mathcal{E}_\mathcal{G}^2 + \mathcal{E}_{c\tau}^2$ which may use the whole tolerance $\mathtt{TOL}_S = 5.0 \cdot 10^{-3}$. Since the coarsen time error $\mathcal{E}_{c\tau}$ is relatively small (cf. Figure 5.8), the fact that AST-FEM exhausts well the tolerance $\mathtt{TOL}_S$, implies that the space error is significantly larger than algorithm CLASSIC allows for, compare Figure 5.9. Comparing the degrees of freedom employed by ASTFEM and CLASSIC depicted in Figure 5.10 (cf. the respective meshes shown in Figures 5.4 and 5.5) we realize that reusing spare tolerance can indeed save a lot of computational effort. This is also reflected in the computational times depicted in Figure 5.10.

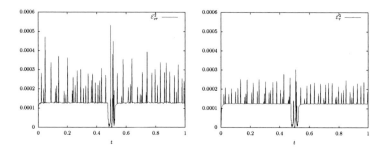

Figure 5.8: Combined coarsen time error in ASTFEM (left) and time error in CLASSIC (right).

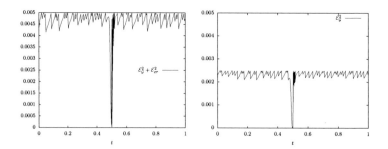

Figure 5.9: Adapted quantity of ASTFEM (left) and space error estimator of CLASSIC (right).

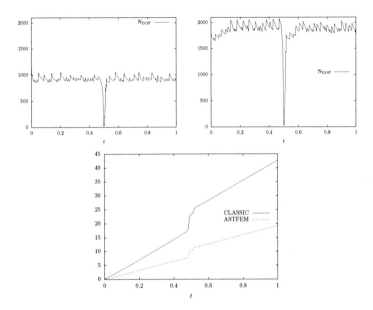

Figure 5.10: Number of degrees of freedom in ASTFEM (top left) and CLASSIC (top right) together with used CPU time in seconds on an AMD Opteron 252 at 2.6GHz with 8GB RAM (bottom) for wandering peak example.

Sinusoidal Gaussian bell

Besides this example, where the spatial adaptivity is employed to capture the movement of a peak, we briefly present another example, where the solution does not exhibit such a movement but rather changes its shape stationary in space. More precisely, we again consider the heat equation on the domain $\Omega = (-1, 1) \times (-1, 1)$ and the time interval $(0, 1)$. This time, however, we use a right hand side such that the exact solution is given by

$$u(x, t) = \sin(\pi t)\, e^{-10\|x\|^2},$$

which can be described as a Gaussian bell in space oscillating sinusoidally in time. We use the same parameters as before and demand the tolerances $\mathtt{TOL}_0 := 2.5 \cdot 10^{-7}$, $\mathtt{TOL}_f := 2.5 \cdot 10^{-4}$ and

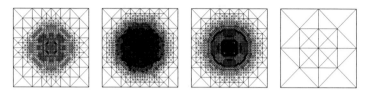

Figure 5.11: Adapted meshes for $t = 0.12, 0.52, 0.748, 0.992$ respectively.

$\text{TOL}_S := 5 \cdot 10^{-4}$. Employing $\text{TOL}_* := 2.5 \cdot 10^{-4}$ leads to a minimal time step size (with respect to the time discretization error) of $\tau_* = 4.12 \cdot 10^{-7}$.

Again, we observe from the meshes depicted in Figure 5.11 that the spatial adaptivity performs well and resolves the transition from $u(\cdot, 0) \equiv 0$ to the full Gaussian bell for $t = 0.5$ and back to $u(\cdot, 1) \equiv 0$ by consecutively refining the mesh from $t = 0$ to $t = 0.5$ before coarsening it again, see also the number of degrees of freedom shown in Figure 5.13. Further, we see from Figure 5.12 that also the time step size is adapted as one expects when considering that the time dependency is given by the factor $\sin(\pi t)$. Considering that the consistency tolerance is given as $\text{TOL}_f := 2.5 \cdot 10^{-4}$, we see from Figure 5.12 and the small coarsen time error indicator revealed by Figure 5.13 that again — as in the previous wandering peak example — the time step size is exclusively determined by the consistency error and in particular, the minimal time step size τ_* is not reached. We point out that even though CONSISTENCY tries to enlarge the time step size in case of $\mathcal{E}_f^2 \leq \sigma\text{TOL}_f = 1.25 \cdot 10^{-4}$, this not necessarily implies that τ is actually enlarged in this case since the enlarged τ might violate the consistency tolerance and thus is reduced again, compare the realization of CONSISTENCY in Algorithm 6. Since this determination of the time step size by the consistency error leads to quite a small coarsen time error (cf. Figure 5.13), ASTFEM again highly profits from the flexibility gained by considering the space and coarsen time error as a pooled quantity.

Finally, in Figure 5.14 we present a close-up of the space and coarsen time error as well as the related number of degrees of freedom, depicting only four time steps of the considered computa-

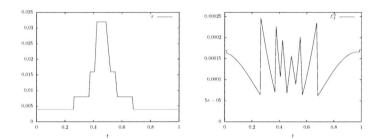

Figure 5.12: Time step sizes and corresponding consistency error produced by ASTFEM for sinusoidal Gaussian bell example. The minimal time step size $\tau_* = 4.12 \cdot 10^{-7}$ is not reached.

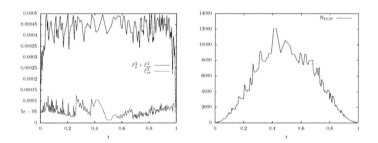

Figure 5.13: Adapted quantity of ASTFEM and combined coarsen time error (left) as well as employed degrees of freedom (right) for sinusoidal Gaussian bell example.

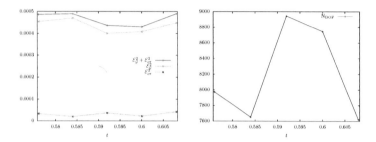

Figure 5.14: Detailed view of space and coarsen time indicator (left) and degrees of freedom (right).

tion. Here, we particularly observe the implications of a carried out adaptation: At time $t = 0.584$, the sum $\mathcal{E}_{\mathcal{G}}^2 + \mathcal{E}_{cT}^2$ (which is the crucial basic quantity for the adaptation in ASTFEM) nearly exhausts the tolerance $\mathsf{TOL_S}$. It is thus not surprising, that for $t = 0.592$ an adaptation was carried out and since the space error dominates the coarsen time error, this comes in the guise of refining the grid; the number of DOFs increases to 8945. As a consequence, the space error $\mathcal{E}_{\mathcal{G}}^2$ (green) drops. However, note that the coarsen time error (blue) increases. This is due to the fact that the new solution (which lives on a fine grid) resolves information of the right hand side that was invisible to the old solution (on the coarse grid). Hence, the spatial refinement increases the time error indicator and consequently also $\mathcal{E}_{cT}^2$ (blue). However, this growth is far too moderate to significantly oppose the reduction of the adapted quantity (red) or to even lead to an increase of $\mathcal{E}_{\mathcal{G}}^2 + \mathcal{E}_{cT}^2$. Such effects, however, can be created by using a very large consistency tolerance, which leads to a coarsen time error being about the same size as the space error.

Even though in the next step ($t = 0.6$), the mesh is slightly coarsened again, the coarsen time error decreases again, which in turn weakens the growth of the sum $\mathcal{E}_{\mathcal{G}}^2 + \mathcal{E}_{cT}^2$.

Chapter 6

Application to carbonation problem

In this chapter, we finally present a real life application for algorithm ASTFEM, namely the adaptive simulation of concrete carbonation, cf. [35]. After providing an overview of the process of concrete carbonation in Section 6.1, we present a homogenized model for concrete carbonation, which was originally derived in [49], in Section 6.2. We anticipate that this model does not fit precisely into the framework of the linear model problem (2.1) which was considered in the previous chapters. Thus, the theoretical results, particularly the a posteriori error estimation of Theorem 3.4.6 and consequently the strict error control of ASTFEM, are not valid for this problem. Nevertheless, in Section 6.3, we derive two different discretizations of the homogenized model and we focus on applying ASTFEM to those discretizations in Section 6.4. Finally, in Section 6.5, we present numerical results and also compare the two discretization approaches.

6.1 Overview of concrete carbonation

Concrete carbonation takes place in the pores of concrete, which are partially saturated with water that clings to the pore walls, and involves reaction, diffusion, precipitation and dissolution. Atmospheric CO_2 enters the concrete through the air-filled pores and dissolves in the pore water. There, it reacts with dissolved constituents of the cement paste, most importantly with $Ca(OH)_2$. This causes a lowering of the pH, facilitating the corrosion of the steel reinforcements and, consequently, leads to a severe reduction of the service life of the structure. The dominant carbonation reaction is usually assumed to be

$$CO_2(aq) + Ca(OH)_2(aq) \longrightarrow CaCO_3(aq) + H_2O. \qquad (6.1)$$

The produced $CaCO_3$ precipitates very quickly to the solid matrix. Detailed surveys on the carbonation problem were carried out, for instance, by [28, 8, 12, 13] and, from a more mathematical point of view, by [37, 38].

Many researchers have tried to find simple formulas for the advancement of the carbonation in a concrete sample, cf. [60, 39] for discussions of these. There are also some works with respect to modeling carbonation with reaction–diffusion systems, most notably [51, 61, 34, 50, 37, 38, 31]. In the homogenization context, it is also worth noting the article [52], where ion diffusion in concrete is modeled using spatial averaging, and [33], where a two-scale model for a simple carbonation scenario is proposed. Here, we want to use the results of [49] to present an efficient numerical (multi-scale) model of concrete carbonation.

An important feature of concrete carbonation is that the carbonation reaction causes a change of the microstructure in at least two ways. First, the reactant $Ca(OH)_2$ takes up considerably less volume than the product $CaCO_3$. This causes a permanent reduction of the pore-air volume. Second, water is produced in the carbonation reaction. This induces a (usually) temporary reduction of the pore-air volume and an increase of the volume of the reaction medium. While it seems that the latter effect is of considerable importance only in accelerated carbonation tests (as opposed to carbonation under natural atmospheric conditions), cf. [39], the first effect is always important because it permanently slows down the diffusion of $CO_2(g)$ to the reaction zone and, in turn, reduces the speed of the overall carbonation process, also cf. [34]. A detailed discussion of these effects and their impact on the progress of carbonation can be found in [39].

It was derived in [49] that a homogenized model for concrete carbonation is as follows,

$$\partial_t J^{\mathrm{a}}(x,t) = -|Z^{\mathrm{w}}| \, C^{\mathrm{m}} C^{\mathrm{H}} R u v^{\mathrm{w}}, \quad x \in \Omega, \ t \in S, \tag{6.2a}$$

$$\partial_t \left((J^{\mathrm{a}} |Z^{\mathrm{a}}| + |Z^{\mathrm{w}}|) u(x,t) \right)$$
$$-\nabla \cdot (J^{\mathrm{a}} D^{\mathrm{a}} P^{\mathrm{a}} \nabla u) = -|Z^{\mathrm{w}}| \, m^{\mathrm{u}} R u v^{\mathrm{w}}, \quad x \in \Omega, \ t \in S, \tag{6.2b}$$

$$|Z^{\mathrm{w}}| \, \partial_t v^{\mathrm{w}}(x,t) = -|Z^{\mathrm{w}}| C^{\mathrm{H}} m^{\mathrm{v}} R u v^{\mathrm{w}}, \quad x \in \Omega, \ t \in S, \tag{6.2c}$$

$$-J^{\mathrm{a}} D^{\mathrm{a}} P^{\mathrm{a}} \nabla u \cdot \eta = C^{\mathrm{ext}} (u - u^{\mathrm{ext}}), \quad x \in \partial\Omega, \ t \in S, \tag{6.2d}$$

where u is a combined concentration of CO_2 in pore air and pore water, v^{w} is the concentration of $Ca(OH)_2$ in pore water and J^{a} describes the reduction of pore-air volume by the carbonation process. The reaction takes place in a spatial domain Ω and on a time interval $S = (0, T)$. The parameters of the model are: $|Z^{\mathrm{a}}|$ – initial pore-air volume fraction, $|Z^{\mathrm{w}}|$ – initial pore-water volume fraction, C^{m} – volume-reduction factor, D^{a} – diffusivity of CO_2 in air, P^{a} – initial tortuosity of pore air, C^{H} – Henry constant, R – reaction constant, m^i, $i \in \{\mathrm{u}, \mathrm{v}\}$ – molar weights, u^{ext} – external CO_2 concentration, C^{ext} – Robin constant.

In the following section, we summarize how system (6.2) is derived by the multi-scale modeling approach in [49].

6.2   Derivation of the macroscopic model

In this section, we derive the system of equations (6.2) from a pore-scale first-principles carbonation model. This can be viewed as a brief but self-consistent summary of the results of [49] directly applied to the problem of concrete carbonation.

The carbonation problem is given in a domain, whose microstructure undergoes an evolution with respect to time. In order to use periodic-homogenization techniques (cf. [7, 53], e.g.), the method of homogenization in domains with evolving microstructure [44] was used in [49] for coping with the changing microstructure. This method is based on transforming the problem from the current configuration to an associated *periodic* reference configuration and to homogenize the resulting transformed problem. In particular, the method allows for non-periodic evolving domains.

It is well-known that scaling of the material parameters with powers of the homogenization parameter has a strong influence on the *type* of limit problem (cf. [47, 42, 41], e.g.). It is therefore of great importance to use the *correct* scaling in order to obtain useful macroscopic models. We introduce a non-dimensionalization in order to obtain problem-dependent scalings, which lead to process-adapted models in the homogenization limit. The particular choice of scaling then depends on the characteristic lengths of the medium as well as the diffusivities and it turns out that the parameters of the concrete-carbonation problem lead to the limit problem (6.2).

6.2.1   The pore-scale carbonation model

We denote the mass concentrations of CO_2 and $Ca(OH)_2$ by $u = u(x,t)$ and $v = v(x,t)$, respectively, where x and t are the space and time variable, respectively, and we distinguish between the concentration in the pore air and the pore water by superscripts a and w, respectively. Note that v is assumed to occur in the pore water only.

We account for two different interfaces in the pores: the air–water interface, $\Gamma^{\mathrm{aw}}(t)$, and the water–solid interface, $\Gamma^{\mathrm{ws}}(t)$. For simplicity, we assume that there is no air–solid interface as no transport has to be accounted for across this interface anyway. For ease of notation, we write $\Gamma^{\mathrm{w}}(t) = \Gamma^{\mathrm{aw}}(t) \cup \Gamma^{\mathrm{ws}}(t)$ and $\Gamma(t) = \Gamma^{\mathrm{aw}}(t)$.

The considered reaction–diffusion problem is then given as follows. The mass balances of the species are

$$\partial_t u^{\mathrm{a}}(x,t) - \nabla \cdot (D^1 \nabla u^{\mathrm{a}} - w^{\mathrm{a}} u^{\mathrm{a}}) = 0, \qquad x \in \Omega^{\mathrm{a}}(t),\ t \in S,$$
(6.3a)

$$\partial_t u^{\mathrm{w}}(x,t) - \nabla \cdot (D^2 \nabla u^{\mathrm{w}} - w^{\mathrm{w}} u^{\mathrm{w}}) = -R^{\mathrm{u}} u^{\mathrm{w}} v^{\mathrm{w}}, \quad x \in \Omega^{\mathrm{w}}(t),\ t \in S,$$
(6.3b)

$$\partial_t v^{\mathrm{w}}(x,t) - \nabla \cdot (D^3 \nabla v^{\mathrm{w}} - w^{\mathrm{w}} v^{\mathrm{w}}) = -R^{\mathrm{v}} u^{\mathrm{w}} v^{\mathrm{w}}, \quad x \in \Omega^{\mathrm{w}}(t),\ t \in S,$$
(6.3c)

where D^i, $i \in \{1,2,3\}$, are the corresponding diffusivities and w^i, $i \in \{\mathrm{a},\mathrm{w}\}$, are the velocities of the deformation of the respective domains. The boundary conditions at the internal boundaries of the porous medium are given by

$$-(D^1 \nabla u^{\mathrm{a}}) \cdot \eta^{\mathrm{a}} = (D^2 \nabla u^{\mathrm{w}}) \cdot \eta^{\mathrm{w}} = C^{\mathrm{ex}} (C^{\mathrm{H}} u^{\mathrm{a}} - u^{\mathrm{w}}),$$
$$x \in \Gamma(t), \ t \in S, \qquad (6.3\mathrm{d})$$
$$-(D^2 \nabla u^{\mathrm{w}}) \cdot \eta^{\mathrm{w}} = 0, \qquad x \in \Gamma^{\mathrm{ws}}(t), \ t \in S, \qquad (6.3\mathrm{e})$$
$$-(D^3 \nabla v^{\mathrm{w}}) \cdot \eta^{\mathrm{w}} = 0, \qquad x \in \Gamma^{\mathrm{w}}(t), \ t \in S, \qquad (6.3\mathrm{f})$$

where η^i denotes the outer normal on $\partial \Omega^i(t)$, $i \in \{\mathrm{a}, \mathrm{w}\}$, the positive Henry constant C^{H} describes the ratio of u^{w} and u^{a} in equilibrium and C^{ex} is related to the rate of interfacial exchange. For now, we assume no-flux conditions at the external interface $\partial \Omega$ for simplicity. The system is completed by non-negative initial conditions, $u^{\mathrm{a}}(x, 0) = u_0^{\mathrm{a}}(x)$, $u^{\mathrm{w}}(x, 0) = u_0^{\mathrm{w}}(x)$ and $v^{\mathrm{w}}(x, 0) = v_0^{\mathrm{w}}(x)$.

Still missing at this stage is a model coupling the reaction–diffusion mechanisms with the evolution of the microstructure. Although this could be done for the microscopic model and carried through the homogenization process as done in [45, 46], we only discuss how the homogenization works for given evolution and model the coupling afterwards. This saves us many technical details.

6.2.2 Non-dimensionalization

In order to obtain process-adapted models in the homogenization limit, it is important to identify the characteristic microscopic and macroscopic lengths and to nondimensionalize the system of equations properly (cf. [48, 41], e.g.). In this section, a possible non-dimensionalization of problem (6.3) is given, which is similar to that given in [47] for a very much simplified version of the problem considered here.

The following dimensionless concentrations are introduced,

$$\tilde{u}^{\mathrm{a}} := u^{\mathrm{a}} C^{\mathrm{H}} / u_{\mathrm{ref}}^{\mathrm{a}}, \qquad \tilde{u}^{\mathrm{w}} := u^{\mathrm{w}} / u_{\mathrm{ref}}^{\mathrm{w}}, \qquad \tilde{v}^{\mathrm{w}} := v^{\mathrm{w}} / v_{\mathrm{ref}}, \qquad (6.4)$$

where u_{ref}^α, $\alpha \in \{\mathrm{a}, \mathrm{w}\}$, and $v_{\mathrm{ref}}^{\mathrm{w}}$ are reference concentrations representing upper bounds on the concentrations. For the carbonation problem, it is natural to take $u_{\mathrm{ref}}^{\mathrm{a}} = u_{\mathrm{ref}}^{\mathrm{w}} =: u_{\mathrm{ref}} = C^{\mathrm{H}} u^{\mathrm{ext}}$ and $v_{\mathrm{ref}} = \|v_0^{\mathrm{w}}\|_\infty$.

Moreover, we introduce a characteristic macroscopic length L (e.g. the diameter of the sample) and a characteristic microscopic length ℓ (e.g. a typical pore diameter). For each concentration u^{a}, u^{w} and v^{w}, we define a characteristic length associated with the diffusion of the species in the pore, ℓ_i^{D}, $i = 1, 2, 3$, respectively, and express this as a multiple of ℓ in the following way: $\ell_i^{\mathrm{D}} = \sqrt{D_i}\ell$ (note the difference between the diffusivity D^i and the number D_i). Letting D_{ref}^i be the L^{∞}-bounds of D^i, $i = 1, 2, 3$, the characteristic diffusion times are then defined as

$$T^{\mathrm{a}} := (\ell_1^{\mathrm{D}})^k L^{2-k}/D_{\mathrm{ref}}^1,$$
$$T^{\mathrm{w}} := (\ell_2^{\mathrm{D}})^l L^{2-l}/D_{\mathrm{ref}}^2,$$
$$T_{\mathrm{v}}^{\mathrm{w}} := (\ell_3^{\mathrm{D}})^j L^{2-j}/D_{\mathrm{ref}}^3. \tag{6.5}$$

Moreover, we set $\varepsilon = \ell/L$.

In order to account for all processes optimally, it is desirable that all processes happen on the same time scale. Therefore, k, l and j need to be chosen such that the characteristic times of diffusion of all species are equal to that of the fastest species, which we choose as close to unity as possible, $T^{\mathrm{a}} = \max\{1, L^2/D_{\mathrm{ref}}^1\}$. Therefore, we set $T^{\mathrm{w}} = T_{\mathrm{v}}^{\mathrm{w}} = T^{\mathrm{a}} =: \tilde{T}$. This determines k, l and j.

Moreover, we introduce the parameter $m \in \mathbb{R}$ related to the speed of the interfacial exchange. The parameter m is to be chosen such that $C^{\mathrm{ex}} \approx \varepsilon^m L/\tilde{T}$, i.e. m needs to be chosen small if C^{ex} is large and it needs to be chosen large if C^{ex} is small.

Making use of the characteristic lengths introduced earlier, let $\tilde{x} := x/L$ and $\tilde{t} := t/\tilde{T}$ be the dimensionless (macroscopic) space and time variables and write $\tilde{u}^{\alpha}(\tilde{x}, \tilde{t}) := u^{\alpha}(\tilde{x}L, \tilde{t}\tilde{T})$, $\alpha \in \{\mathrm{a}, \mathrm{w}\}$, and, analogously, for v^{w}. Note that the corresponding time interval is then given by $\tilde{S} := S/\tilde{T}$.

Introducing the following dimensionless variables and parameters,

$$\tilde{w}^{\mathrm{a}} := \frac{\tilde{T}}{L} w^{\mathrm{a}}, \quad \tilde{w}^{\mathrm{w}} := \frac{\tilde{T}}{L} w^{\mathrm{w}}, \tag{6.6}$$

$$\tilde{D}^{\mathrm{a}} := \frac{D_1^{\frac{k}{2}}}{D_{\mathrm{ref}}^1} D^1, \quad \tilde{D}^{\mathrm{w}} := \frac{D_2^{\frac{l}{2}}}{D_{\mathrm{ref}}^2} D^2, \quad \tilde{E}^{\mathrm{w}} := \frac{D_3^{\frac{j}{2}}}{D_{\mathrm{ref}}^3} D^3, \tag{6.7}$$

$$\tilde{a} := \ell^{-m} L^{m-1} \tilde{T} C^{\mathrm{ex}}, \tag{6.8}$$

$$\tilde{R}^{\mathrm{u}} := \tilde{T} u_{\mathrm{ref}} R^{\mathrm{u}}, \quad \tilde{R}^{\mathrm{v}} := \tilde{T} v_{\mathrm{ref}} R^{\mathrm{v}}, \tag{6.9}$$

the dimensionless version of system (6.3) is given by

$$\partial_{\tilde{t}}\tilde{u}^{\mathrm{a}} - \nabla \cdot (\varepsilon^k \tilde{D}^{\mathrm{a}} \nabla \tilde{u}^{\mathrm{a}} - \tilde{w}^{\mathrm{a}}\tilde{u}^{\mathrm{a}}) = 0, \qquad \tilde{x} \in \tilde{\Omega}^{\mathrm{a}}(t), \ t \in \tilde{S},$$

$$\partial_{\tilde{t}}\tilde{u}^{\mathrm{w}} - \nabla \cdot (\varepsilon^l \tilde{D}^{\mathrm{w}} \nabla \tilde{u}^{\mathrm{w}} - \tilde{w}^{\mathrm{w}}\tilde{u}^{\mathrm{w}}) = -\tilde{R}^{\mathrm{u}}\tilde{u}^{\mathrm{w}}\tilde{v}^{\mathrm{w}}, \qquad \tilde{x} \in \tilde{\Omega}^{\mathrm{w}}(t), \ t \in \tilde{S},$$

$$\partial_{\tilde{t}}\tilde{v}^{\mathrm{w}} - \nabla \cdot (\varepsilon^j \tilde{E}^{\mathrm{w}} \nabla \tilde{v}^{\mathrm{w}} - \tilde{w}^{\mathrm{w}}\tilde{v}^{\mathrm{w}}) = -\tilde{R}^{\mathrm{v}}\tilde{u}^{\mathrm{w}}\tilde{v}^{\mathrm{w}}, \qquad \tilde{x} \in \tilde{\Omega}^{\mathrm{w}}(t), \ t \in \tilde{S},$$

$$\tag{6.10a}$$

$$-\varepsilon^k \tilde{D}^{\mathrm{a}} \nabla \tilde{u}^{\mathrm{a}} \cdot \tilde{\eta}^{\mathrm{a}} = \varepsilon^m \tilde{a} C^{\mathrm{H}} (\tilde{u}^{\mathrm{a}} - \tilde{u}^{\mathrm{w}}), \qquad \tilde{x} \in \tilde{\Gamma}(t), \ t \in \tilde{S},$$

$$-\varepsilon^l \tilde{D}^{\mathrm{w}} \nabla \tilde{u}^{\mathrm{w}} \cdot \tilde{\eta}^{\mathrm{w}} = -\varepsilon^m \tilde{a}(\tilde{u}^{\mathrm{a}} - \tilde{u}^{\mathrm{w}}), \qquad \tilde{x} \in \tilde{\Gamma}(t), \ t \in \tilde{S},$$

$$-\varepsilon^l \tilde{D}^{\mathrm{w}} \nabla \tilde{u}^{\mathrm{w}} \cdot \tilde{\eta}^{\mathrm{w}} = 0, \qquad \tilde{x} \in \tilde{\Gamma}^{\mathrm{ws}}(t), \ t \in \tilde{S},$$

$$-\varepsilon^j \tilde{E}^{\mathrm{w}} \nabla \tilde{v}^{\mathrm{w}} \cdot \tilde{\eta}^{\mathrm{w}} = 0, \qquad \tilde{x} \in \tilde{\Gamma}^{\mathrm{w}}(t), \ t \in \tilde{S}.$$

$$\tag{6.10b}$$

Note that no assumption on the periodicity of the domains has been made so far but only the existence of macroscopic and microscopic characteristic lengths, i.e. a separation of distinct scales.

6.2.3 Transformation to an ε-periodic reference configuration

In order to apply (periodic) homogenization techniques, the *method of homogenization in domains with evolving microstructure* [44] can be used. It is classically not possible to account for an evolution of the microstructure using periodic homogenization, in particular if the evolution is different in (macroscopically) different places of the medium. The idea of the method employed is to consider the problem in a fixed reference geometry and use a time- and space-dependent mapping, which accounts for the evolution of the microstructure. In terms of classical continuum mechanics, this corresponds to transforming the system from being stated in the current configuration (or Eulerian description; such as (6.10)) to an appropriate reference configuration (or Lagrangian description). If the reference geometry can be chosen periodic, then periodic-homogenization ideas can be applied. The method is particularly suitable for problems, where the evolution of the microstructure is induced by the chemical mechanism itself [45].

At $t = 0$, the heterogeneous material is assumed to be periodic with respect to a reference cell $Y = (0,1)^n$ scaled by a (small) scale parameter ε. This assumption can be relaxed — it suffices if there exists a periodic reference configuration onto which the initial geometry can be mapped by a diffeomorphism — but it simplifies the presentation considerably. The reference cell contains a solid particle, Z^s, some pore water, Z^w, and void (pore) space Z^a (each being an open bounded domain with Lipschitz-continuous boundary), i.e. at $t = 0$ the domain Ω is the union of a finite amount of translated versions of εY. Then, $\Omega_\varepsilon^\alpha(0) = \Omega \cap \text{int} \bigcup_k \varepsilon \overline{Z}_k^\alpha$, $\alpha \in \{a, w, s\}$, where the subscript k denotes translation of the set by $k \in \mathbb{Z}^n$ and ε indicates the ε-periodic geometry of the domain. The evolution of the parts of Ω is assumed to be such that it can be described by orientation-preserving mappings $\psi_\varepsilon^i(\,\cdot\,, t) \colon \Omega_\varepsilon^i \to \Omega^i(t)$, $i \in \{a, w, s\}$, for each $t \in S$, whose images are the current spatial extents of pore air, pore water and solid matrix, respectively. In cases where the evolution of the domain is a deformation, the corresponding function $\psi_\varepsilon^i(x, \,\cdot\,)$ gives the trajectory of the particle residing at x at time $t = 0$. The idea is not restricted to deformations, however, and can be used to model addition or removal of substance (increase of volume of solid matrix by precipitation e.g.), cf. [44].

We define $\Psi_\varepsilon^i = \nabla \psi_\varepsilon^i$ and $J_\varepsilon^i = \det \Psi_\varepsilon^i$, which we will require to be bounded from above and away from zero. In particular, these assumptions restrict the evolutions to ones where no changes of the topology of the subdomains occur and where all subdomains always have positive volume. The advantage of this construction is that the parts of Ω may evolve differently in different places. The problem in the reference configuration is posed on an ε-periodic medium, however, and is hence suitable for an analysis in the context of periodic homogenization.

The idea of periodic homogenization cf. e. g., [7, 53] is then to examine the limit as ε approaches zero in order to obtain averaged problems defined in all of Ω which are easier to treat numerically and give useful information about macroscopically observable processes. The unknowns in the ε-periodic reference configuration are denoted by u_ε^a, u_ε^w and v_ε^w. In accordance with (6.10), we consider the model equations (plus initial conditions and homogeneous Neumann conditions at the exterior boundary for all concentrations) in the reference configuration. This can be obtained straightforwardly using some of the calculus of mechanics of deformable bodies and is given by (cf. [44])

$$\partial_t(J_\varepsilon^a u_\varepsilon^a) - \nabla \cdot (\varepsilon^k J_\varepsilon^a \Psi_\varepsilon^{a-1} D_\varepsilon^a \Psi_\varepsilon^{a-T} \nabla u_\varepsilon^a) = 0, \qquad x \in \Omega_\varepsilon^a, \ t \in S,$$

$$\partial_t(J_\varepsilon^w u_\varepsilon^w) - \nabla \cdot (\varepsilon^l J_\varepsilon^w \Psi_\varepsilon^{w-1} D_\varepsilon^w \Psi_\varepsilon^{w-T} \nabla u_\varepsilon^w) = -R_\varepsilon^u J_\varepsilon^w u_\varepsilon^w v_\varepsilon^w,$$
$$x \in \Omega_\varepsilon^w, \ t \in S,$$

$$\partial_t(J_\varepsilon^w v_\varepsilon^w) - \nabla \cdot (\varepsilon^j J_\varepsilon^w \Psi_\varepsilon^{w-1} E_\varepsilon^w \Psi_\varepsilon^{w-T} \nabla v_\varepsilon^w) = -R_\varepsilon^v J_\varepsilon^w u_\varepsilon^w v_\varepsilon^w,$$
$$x \in \Omega_\varepsilon^w, \ t \in S,$$
$$\text{(6.11a)}$$

$$-(\varepsilon^k J_\varepsilon^a \Psi_\varepsilon^{a-1} D_\varepsilon^a \Psi_\varepsilon^{a-T} \nabla u_\varepsilon^a) \cdot \eta_\varepsilon^a = \varepsilon^m a_\varepsilon C^H \left\| \Psi_\varepsilon^{a-T} \eta_\varepsilon^a \right\| J_\varepsilon^a (u_\varepsilon^a - u_\varepsilon^w),$$
$$x \in \Gamma_\varepsilon, \ t \in S,$$

$$-(\varepsilon^l J_\varepsilon^w \Psi_\varepsilon^{w-1} D_\varepsilon^w \Psi_\varepsilon^{w-T} \nabla u_\varepsilon^w) \cdot \eta_\varepsilon^w = -\varepsilon^m a_\varepsilon \left\| \Psi_\varepsilon^{w-T} \eta_\varepsilon^w \right\| J_\varepsilon^w (u_\varepsilon^a - u_\varepsilon^w),$$
$$x \in \Gamma_\varepsilon, \ t \in S,$$

$$-(\varepsilon^l J_\varepsilon^w \Psi_\varepsilon^{w-1} D_\varepsilon^w \Psi_\varepsilon^{w-T} \nabla u_\varepsilon^w) \cdot \eta_\varepsilon^w = 0, \qquad x \in \Gamma_\varepsilon^{ws}, \ t \in S,$$

$$-(\varepsilon^j J_\varepsilon^w \Psi_\varepsilon^{w-1} E_\varepsilon^w \Psi_\varepsilon^{w-T} \nabla v_\varepsilon^w) \cdot \eta_\varepsilon^w = 0, \qquad x \in \Gamma_\varepsilon^w, \ t \in S,$$
$$\text{(6.11b)}$$

where quantities with a subscript ε in the reference configuration correspond to the same quantities with the tilde in the current configuration, $\tilde{x} = \psi_\varepsilon^a(x, t)$.

A very important observation is that the functions ψ_ε^i actually do not appear in (6.11) but only Ψ_ε^i and J_ε^i. We will make use of this fact when coupling the evolution to the reaction–diffusion process modeling concrete carbonation in Section 6.2.4.

The homogenized limit problem strongly depends on the values of the scaling parameters k, l, j and m. Therefore, these need to be calculated before passing to the limit. The typical capillary pore diameter in Ordinary Portland Cement based concrete can vary between 10^{-7} cm and 10^{-4} cm, cf. [43]. For the present simulations, we take $\ell = 10^{-4}$ cm, $\ell_1^D = 1$ and $\ell_2^D = \ell_3^D = \ell/\sqrt{625}$ as the characteristic microscopic lengths. Note that ℓ_2^D and ℓ_3^D are essentially smaller than ℓ owing to the thinness of the water film in concrete pores. Moreover, we choose $L = 1$cm. This implies $\varepsilon = 10^{-4}$. The diffusion coefficient of $CO_2(g)$ is $D^1 = 13.8 \cdot 10^3$ cm^2/d and diffusivities of the ions in water are $D^2 = D^3 = 1$ cm^2/d, cf. [29]. The data given above yields characteristic times of

$$T^{\mathrm{a}} = 10^{-4k}1^{2-k}/13.8 \cdot 10^3,$$

$$T^{\mathrm{w}} = 25^l 10^{-4l}1^{2-l},$$

$$T^{\mathrm{w}}_{\mathrm{v}} = 25^j 10^{-4j}1^{2-j}. \tag{6.12}$$

According to Section 6.2.2, this implies $k = 0$, $l = j \approx 1.59$ and $T^{\mathrm{a}} = T^{\mathrm{w}} = T^{\mathrm{w}}_{\mathrm{v}} = 1/13.8 \cdot 10^3$. For the interfacial–exchange coefficient at the internal gas–liquid interface, we use the empirical value $C^{\mathrm{ext}} = 10^3/\mathrm{d}$. Aiming for $\tilde{a} \approx 1$, this implies for the scaling exponent $m \approx 0.29$.

For these choices of scaling exponents, the homogenized system as $\varepsilon \to 0$ is given by (cf. [49])

$$\partial_t((J^{\mathrm{a}}\,|Z^{\mathrm{a}}| + J^{\mathrm{w}}\,|Z^{\mathrm{w}}|)u(x,t))$$
$$-\nabla \cdot (J^{\mathrm{a}} D^{\mathrm{a}} P^{\mathrm{a}} \nabla u) = -J^{\mathrm{w}}\,|Z^{\mathrm{w}}|\,R^{\mathrm{u}}uv^{\mathrm{w}}, \qquad x \in \Omega,\ t \in S, \tag{6.13a}$$

$$|Z^{\mathrm{w}}|\,\partial_t(J^{\mathrm{w}}v^{\mathrm{w}}(x,t)) = -J^{\mathrm{w}}|Z^{\mathrm{w}}|C^{\mathrm{H}}R^{\mathrm{v}}uv^{\mathrm{w}}, \qquad x \in \Omega,\ t \in S, \tag{6.13b}$$

$$-J^{\mathrm{a}} D^{\mathrm{a}} P^{\mathrm{a}} \nabla u \cdot \eta = C^{\mathrm{ext}}(u - u^{\mathrm{ext}}), \qquad x \in \partial\Omega,\ t \in S. \tag{6.13c}$$

Note that $u = C^{\mathrm{H}}u^{\mathrm{a}} = u^{\mathrm{w}}$ represents a *combined* CO_2 concentration referring to both pore air and pore water, where $J^{\mathrm{a}} D^{\mathrm{a}} P^{\mathrm{a}}$ is the effective diffusion tensor.

As we are only interested in cases, where $J^{\mathrm{a}} \Psi^{\mathrm{a}-1} D^{\mathrm{a}} \Psi^{\mathrm{a}-T}$ is solely dependent on the macro-variable, it is convenient to define the effective diffusion tensor without an evolution of the microstructure and with unit diffusivity first. This can be viewed as the initial tortuosity of the porous material and it is denoted by Q. The elements of this symmetric and positive definite tensor (e.g. cf. [16]) are given by

$$q_{jk} = \frac{1}{|Z^{\mathrm{a}}|} \int_{Z^{\mathrm{a}}} (\delta_{jk} + \partial_{y_j}\varsigma_k(y))\, dy, \tag{6.14}$$

where δ_{jk} is the Kronecker symbol and ς_j, $j = 1,\ldots,n$, are the Y-periodic solution of the cell problem

$$-\nabla_y \cdot (\nabla_y\varsigma_j(y) + e_j) = 0, \qquad y \in Z^{\mathrm{a}}, \tag{6.15a}$$

$$-(\nabla_y\varsigma_j(y) + e_j) \cdot \eta^{\mathrm{a}} = 0, \qquad y \in \partial Z^{\mathrm{a}}\backslash\partial Y, \tag{6.15b}$$

the weak form of which is given by

$$\int_{Z^a} (\nabla \varsigma_j + e_j) \nabla \varphi \, dy = 0 \qquad (6.16)$$

for all Y-periodic test functions φ. The macroscopic effective diffusion tensor is then given by

$$P^a = |Z^a| \, \Psi^{a-1} \Psi^{a-T} Q. \qquad (6.17)$$

6.2.4 Coupling the reaction–diffusion system and the evolution of the microstructure

Before solving the system of equations (6.13), we must specify the functions related to the evolution of the subdomains, i.e. J^a, Ψ^a and J^w (Ψ^a is contained in the definition of P^a, cf. (6.17)). For this, recall that Ψ^i relates the length and orientation of a material fiber in the reference configuration to its length and orientation in the current configuration and J^i describes the change of volume, $i \in \{a, w\}$. Since water can be assumed incompressible, it is reasonable to set $J^w = 1$. Diffusion of CO_2 in air is fast; hence, we assume that the influence of the microstructure on the direction of the flux is negligible, i.e. $\Psi^a \approx 1$ in the definition of (6.17). We are left with deriving an equation for J^a, which must be coupled to the reaction since the change of volume of pore air is due to the conversion of $Ca(OH)_2$ to $CaCO_3$.

The dissolution of $Ca(OH)_2$ and precipitation of $CaCO_3$ are fast which is why we approximate them as being instantaneous. In the cell Y at the point x, the amount (i.e., mass) of $Ca(OH)_2$ being used up and of $CaCO_3$ being produced at time t is given by

$$\int_{Z^w} m^v J^w C^H R u^w v^w \, dy \quad \text{and} \quad \int_{Z^w} m^C J^w C^H R u^w v^w \, dy,$$
$$(6.18)$$

respectively, where m^C is the molar weight of $CaCO_3$. Noting that $J^w = 1$ and that an increase of total volume of constituents implies a decrease of volume of pore air, we have

$$\frac{\mathrm{d}}{\mathrm{d}t} \int_{Z^a} J^a \, dy = -|Z^a| \, C^m C^H R \int_{Z^w} u^w v^w \, dy \qquad (6.19)$$

$$\text{where} \quad C^m = \frac{1}{|Z^a|} \left(\frac{m^C}{\rho^C} - \frac{m^v}{\rho^v} \right)$$

for each instant in time, where ρ^{v} and ρ^{C} are the densities of $\mathrm{Ca(OH)_2}$ and $\mathrm{CaCO_3}$, respectively, and m^i, $i \in \{\mathrm{v}, \mathrm{C}\}$, are the molar weights.

Since we do not expect J^{a} to vary within one cell (and since we do not have any more information), we *define* $J^{\mathrm{a}}(x, t)$ to be the constant value determined by (6.19) in each cell. Hence, an equation for J^{a} is obtained: Find $J^{\mathrm{a}} \colon \Omega \times S \to \mathbb{R}$ such that

$$\partial_t J^{\mathrm{a}}(x, t) = -|Z^{\mathrm{w}}| C^{\mathrm{m}} C^{\mathrm{H}} R u^{\mathrm{w}} v^{\mathrm{w}}, \quad J^{\mathrm{a}}(0) \equiv 1. \tag{6.20}$$

Note that no assumption on the specific pore geometry has been made in the derivation of (6.20). Making use of the equality $u = u^{\mathrm{w}}$ and stating the variables explicitly, (6.20) can be written as

$$\partial_t J^{\mathrm{a}}(x, t) = -|Z^{\mathrm{w}}| C^{\mathrm{m}} C^{\mathrm{H}} R u(x, t) \, v^{\mathrm{w}}(x, t)$$
$$x \in \Omega, \ t \in S. \tag{6.21a}$$

The rest of the limit problem remains

$$\partial_t((J^{\mathrm{a}}(x, t) \,|Z^{\mathrm{a}}| + |Z^{\mathrm{w}}|)u(x, t))$$
$$-\nabla \cdot (J^{\mathrm{a}}(x, t) D^{\mathrm{a}} P^{\mathrm{a}} \nabla u(x, t)) = -|Z^{\mathrm{w}}| \, m^{\mathrm{u}} R u(x, t) \, v^{\mathrm{w}}(x, t),$$
$$x \in \Omega, \ t \in S, \tag{6.21b}$$

$$|Z^{\mathrm{w}}| \, \partial_t v^{\mathrm{w}}(x, t) = -|Z^{\mathrm{w}}| \, C^{\mathrm{H}} m^{\mathrm{v}} R u(x, t) \, v^{\mathrm{w}}(x, t),$$
$$x \in \Omega, \ t \in S, \tag{6.21c}$$

$$-J^{\mathrm{a}}(x, t) D^{\mathrm{a}} P^{\mathrm{a}} \nabla u(x, t) \cdot \eta = C^{\mathrm{ext}}(u(x, t) - u^{\mathrm{ext}}),$$
$$x \in \partial\Omega, \ t \in S, \tag{6.21d}$$

where we have explicitly spelled out $R^{\mathrm{u}} = m^{\mathrm{u}} R$ and $R^{\mathrm{v}} = m^{\mathrm{v}} R$. This system is considerably simpler that the original microscopic system. Nevertheless, the correct scaling ensures that all processes are captured accurately in this limit model. The evolution of the microstructure is accounted for by the coupled nature of the equation for the pore-volume factor J^{a} (6.21a) with the reaction–diffusion system (6.21b)–(6.21d).

6.3   Discretization

In the following, we will not go into existence and uniqueness of a solution $u, v^{\mathrm{w}}, J^{\mathrm{a}}$ of problem (6.21), but rather motivate two

slightly different discretizations of the problem and review computational results of these discretizations in Section 6.5. For that pur-
pose, we *assume* that there is a unique solution $u, v^{\mathrm{w}}, J^{\mathrm{a}}$ of (6.21).
As in Section 2, we identify the space and time dependent functions

$$u, v^{\mathrm{w}}, J^{\mathrm{a}} : \Omega \times (0, T) \to \mathbb{R}$$

with functions mapping from a time interval to a Hilbert space of
functions over Ω, particularly

$$u, v^{\mathrm{w}}, J^{\mathrm{a}} : (0, T) \to H^1(\Omega).$$

To allow for convenient notation in the following sections, we treat
the functions $u, v^{\mathrm{w}}, J^{\mathrm{a}}$ as smooth with respect to time and continuous with respect to space, i. e., $u, v^{\mathrm{w}}, J^{\mathrm{a}} \in C^1(0, T; H^1(\Omega) \cap C^0(\Omega))$.
We understand the product $f(t)g(t)$ of two such functions in a
pointwise sense, i. e., $f(t)g(t) : (0, T) \to H^1(\Omega)$ is the identification
of $f(x, t)g(x, t)$.

We note, that problem (6.21) consists of a partial differential
equation (6.21b) with Robin boundary condition (6.21d) which is
coupled with two ordinary differential equations given in (6.21a)
and (6.21c). Moreover, we observe that this coupling is non-linear
since both ODEs and the PDE involve products of u, v^{w}, and J^{a}.
In the following, we present two discretizations of problem (6.21),
which both employ a time stepping through the interval $S = (0, T)$
and an implicit Euler Galerkin discretization of the PDE. In Section 6.3.3, we use the implicit Euler scheme to discretize the ODEs
and together with the implicit Euler Galerkin discretization of the
PDE, this leads to a finite dimensional, non-linear problem, which
we cope with by employing Newton's method. In Section 6.3.4 on
the other hand, we enforce a decoupling of the PDE and the two
ODEs in every time step. This enables us to employ a black box
ODE solver for individually solving the ODEs (6.21a) and (6.21c)
before using their solutions as parameters in the PDE and solving
it as well.

6.3.1   Weak formulation of the PDE

Similar to the derivation of a weak formulation of the linear parabolic equation in Section 2.1, we obtain the weak formulation of

this PDE by multiplying (6.21b) for a fixed time $t \in (0, T)$ with an arbitrary $\varphi \in H^1(\Omega)$ and integrating over Ω. This gives

$$\left\langle \frac{d}{dt} \left((J^{\mathrm{a}}(t) |Z^{\mathrm{a}}| + |Z^{\mathrm{w}}|) u(t) \right), \varphi \right\rangle \tag{6.22}$$

$$- \left\langle \nabla \cdot (J^{\mathrm{a}}(t) D^{\mathrm{a}} P^{\mathrm{a}} \nabla u(t)), \varphi \right\rangle + \left\langle |Z^{\mathrm{w}}| m^{\mathrm{u}} R u(t) v^{\mathrm{w}}(t), \varphi \right\rangle = 0$$

for all $\varphi \in H^1(\Omega)$. However, opposed to the parabolic equation given in (2.1) on page 13, equation (6.21) does not possess (homogeneous) Dirichlet boundary conditions and hence, we employ test functions from the Hilbert space $H^1(\Omega)$ instead of $H^1_0(\Omega)$. Keeping this in mind, we integrate by parts the second order term of equation (6.22) to find

$$-\left\langle \nabla \cdot (J^{\mathrm{a}}(t) D^{\mathrm{a}} P^{\mathrm{a}} \nabla u(t)), \varphi \right\rangle = \left\langle J^{\mathrm{a}}(t) D^{\mathrm{a}} P^{\mathrm{a}} \nabla u(t), \nabla \varphi \right\rangle$$

$$- \int_{\partial \Omega} J^{\mathrm{a}}(t) D^{\mathrm{a}} P^{\mathrm{a}} \nabla u(t) \cdot \eta \, \varphi \, ds,$$

where η denotes the outer unit normal of $\partial \Omega$. Considering the Robin boundary condition of equation (6.21d), this further yields

$$= \left\langle J^{\mathrm{a}}(t) D^{\mathrm{a}} P^{\mathrm{a}} \nabla u(t), \nabla \varphi \right\rangle + C^{\mathrm{ext}} \int_{\partial \Omega} (u(t) - u^{\mathrm{ext}}) \, \varphi \, ds. \tag{6.23}$$

Substituting (6.23) into (6.22), we find the weak formulation of the PDE (6.21b, 6.21d) as

$$\left\langle \frac{d}{dt} \left((J^{\mathrm{a}}(t) |Z^{\mathrm{a}}| + |Z^{\mathrm{w}}|) u(t) \right), \varphi \right\rangle + \left\langle J^{\mathrm{a}}(t) D^{\mathrm{a}} P^{\mathrm{a}} \nabla u(t), \nabla \varphi \right\rangle$$

$$+ \left\langle |Z^{\mathrm{w}}| m^{\mathrm{u}} R u(t) v^{\mathrm{w}}(t), \varphi \right\rangle + C^{\mathrm{ext}} \langle u(t) - u^{\mathrm{ext}}, \varphi \rangle_{\partial \Omega} = 0$$

$$\text{for all } \varphi \in H^1(\Omega). \tag{6.24}$$

Before discretizing this weak formulation, we note that the first term of (6.24) contains the temporal derivative of the product of the two time dependent functions J^{a} and u. We separate this into pure derivatives of either J^{a} or u by employing the product rule.

We rewrite

$$\left\langle \frac{d}{dt}\left((J^{\mathrm{a}}(t)\,|Z^{\mathrm{a}}| + |Z^{\mathrm{w}}|)u(t)\right),\,\varphi\right\rangle = \left\langle |Z^{\mathrm{a}}|\,\frac{d}{dt}\left(J^{\mathrm{a}}(t)\,u(t)\right),\,\varphi\right\rangle$$
$$+ \left\langle |Z^{\mathrm{w}}|\,u'(t),\,\varphi\right\rangle$$
$$= \left\langle (J^{\mathrm{a}}\,|Z^{\mathrm{a}}| + |Z^{\mathrm{w}}|)\,u'(t),\,\varphi\right\rangle + \left\langle |Z^{\mathrm{a}}|\,u(t)\,J^{\mathrm{a}\prime}(t),\,\varphi\right\rangle,$$

and substituting this into equation (6.23) we obtain

$$\left\langle (J^{\mathrm{a}}(t)\,|Z^{\mathrm{a}}| + |Z^{\mathrm{w}}|)\,u'(t),\,\varphi\right\rangle + \left\langle J^{\mathrm{a}}(t)D^{\mathrm{a}}P^{\mathrm{a}}\nabla u(t),\,\nabla\varphi\right\rangle$$
$$+ \left\langle (|Z^{\mathrm{w}}|\,m^{\mathrm{u}}Rv^{\mathrm{w}}(t) + |Z^{\mathrm{a}}|\,J^{\mathrm{a}\prime}(t))\,u(t),\,\varphi\right\rangle$$
$$+ C^{\mathrm{ext}}\langle u(t) - u^{\mathrm{ext}},\,\varphi\rangle_{\partial\Omega} = 0. \tag{6.25}$$

As in the discretization of the linear parabolic equation in Section 3.2, this weak formulation will be the foundation of the discretization of the PDE.

6.3.2 Objective of the discretization

Besides deriving a discretization for the PDE from (6.25), we also have to discretize the two ODEs and in the following, we set up a common setting for the whole discretization process.

For temporal discretization, we divide the time interval $S = (0, T)$ into subintervals $I_n := [t_{n-1}, t_n]$, $n = 1, \ldots, N$, with $0 = t_0 < t_1 < \cdots < t_N = T$ and time step sizes $\tau_n = t_n - t_{n-1}$. With regard to space, we resort to a finite dimensional subspace $\mathbb{V}_n \subset H^1(\Omega)$ in each time step n. We particularly emphasize that those spaces are not nested in general, i.e., $\mathbb{V}_{n-1} \not\subset \mathbb{V}_n$. In the implementation underlying the computational results presented in Section 6.5, we employ finite element spaces $\mathbb{V}_n := \widetilde{\mathbb{V}}(\mathcal{G}_n)$, where $\{\mathcal{G}_n\}_{n=1,\ldots,N}$ are adaptively generated triangulations of Ω. We explicitly point out that in this chapter, we utilize finite element spaces *without* zero boundary, cf. Section 3.1.2.

For given $U_0, V_0, J_0 \in \mathbb{V}_0$, the objective of the considered discretization then is to produce sequences of functions $\{U_n\}_{n=1,\ldots,N}$, $\{V_n\}_{n=1,\ldots,N}$, and $\{J_n\}_{n=1,\ldots,N}$, with $U_n, V_n, J_n \in \mathbb{V}_n$ for all $n = 1, \ldots, N$, where we consider U_n, V_n, and J_n as approximations of $u(t_n)$, $v^{\mathrm{w}}(t_n)$, and $J^{\mathrm{a}}(t_n)$ respectively. Based on these sequences, we introduce the continuous functions U, V, and J via linear interpolation, i. e., we define $U(t)$ as

$$U(t) := \frac{t_n - t}{\tau_n} U_{n-1} + \frac{t - t_{n-1}}{\tau_n} U_n \qquad \text{for } t \in (t_{n-1}, t_n]$$

and $V(t)$ and $J(t)$ are defined accordingly. We consider $U(t)$, $V(t)$, and $J(t)$ as approximations for $u(t)$, $v^{\mathrm{w}}(t)$, and $J^{\mathrm{a}}(t)$ respectively.

In this setting, the implicit Euler Galerkin discretization of equation (6.25) reads: Find $U_n \in \mathbb{V}_n$ such that

$$\left\langle (J_n |Z^{\mathrm{a}}| + |Z^{\mathrm{w}}|) \frac{U_n - U_{n-1}}{\tau_n} , \varphi \right\rangle + \langle J_n D^{\mathrm{a}} P^{\mathrm{a}} \nabla U_n , \nabla \varphi \rangle$$
$$+ \langle (|Z^{\mathrm{w}}| m^{\mathrm{u}} R V_n + |Z^{\mathrm{a}}| J'(t_n)) U_n , \varphi \rangle$$
$$+ C^{\mathrm{ext}} \langle U_n - u^{\mathrm{ext}} , \varphi \rangle_{\partial\Omega} = 0$$
$$\tag{6.26}$$

for all $\varphi \in \mathbb{V}_n$. We observe, that the definition of $J(t)$ as linear interpolation implies $J'(t_n) = (J_n - J_{n-1})/\tau_n$.

Whereas we use this Euler Galerkin discretization of the PDE in both discretizations presented in Sections 6.3.3 and 6.3.4, we will discretize the ODEs (6.21a) and (6.21c) differently in each of those sections. Before proceeding with these discretizations, we first reformulate (6.21a) and (6.21c) by stating them on each time interval I_n individually, i. e.,

$$\frac{d}{dt} J^{\mathrm{a}}(t) = - |Z^{\mathrm{w}}| C^{\mathrm{m}} R C^{\mathrm{H}} u(t) v^{\mathrm{w}}(t) \qquad \text{on } [t_{n-1}, t_n], \quad (6.27a)$$
$$\frac{d}{dt} v^{\mathrm{w}}(t) = -C^{\mathrm{H}} m^{\mathrm{v}} R u(t) v^{\mathrm{w}}(t) \qquad \text{on } [t_{n-1}, t_n], \quad (6.27b)$$

for $n = 1, \ldots, N$. Note that in this formulation, on each time interval I_n, $n \geq 2$, the initial values $J^{\mathrm{a}}(t_{n-1})$ respectively $v^{\mathrm{w}}(t_{n-1})$ are given implicitly by the solution on the previous time interval. Only the values for $J^{\mathrm{a}}(0)$ and $v^{\mathrm{w}}(0)$ must be given explicitly.

6.3.3 Coupled discretization

As a next step, we aim at resorting to a subspace $\mathbb{V}_n \subset H^1(\Omega)$ on each time interval I_n. We cannot do this by straightforwardly considering (6.27a) and (6.27b) and replacing $J^{\mathrm{a}}(t)$ and $v^{\mathrm{w}}(t)$ by functions $J^n(t) \in \mathbb{V}_n$ respectively $V^n(t) \in \mathbb{V}_n$, since $u(t) \notin \mathbb{V}_n$ is involved in both equations. To overcome this and to avoid difficulties arising from the fact that the spaces $\mathbb{V}_{n-1} \not\subset \mathbb{V}_n$ are not nested, we multiply (6.27a) and (6.27b) by an arbitrary test function $\varphi \in H^1(\Omega)$ and integrate over Ω to find

$$\langle \frac{d}{dt} J^{\mathrm{a}}(t) \,, \varphi \rangle = -\langle |Z^{\mathrm{w}}| \, C^{\mathrm{m}} R C^{\mathrm{H}} u(t) \, v^{\mathrm{w}}(t) \,, \varphi \rangle \quad \text{on } [t_{n-1}, t_n],$$

$$\langle \frac{d}{dt} v^{\mathrm{w}}(t) \,, \varphi \rangle = -\langle C^{\mathrm{H}} m^{\mathrm{v}} R u(t) \, v^{\mathrm{w}}(t) \,, \varphi \rangle \quad \text{on } [t_{n-1}, t_n],$$

for all $\varphi \in H^1(\Omega)$. We easily transfer these equations to a subspace $\mathbb{V}_n$ in each time step n by resorting to test functions $\varphi \in \mathbb{V}_n$ and considering the problem: Find $V^n(t), J^n(t) \in \mathbb{V}_n$ such that

$$\langle \frac{d}{dt} J^n(t) \,, \varphi \rangle = -\langle |Z^{\mathrm{w}}| \, C^{\mathrm{m}} R C^{\mathrm{H}} U(t) \, V^n(t) \,, \varphi \rangle \quad \text{on } [t_{n-1}, t_n],$$
$$\tag{6.29a}$$

$$\langle \frac{d}{dt} V^n(t) \,, \varphi \rangle = -\langle C^{\mathrm{H}} m^{\mathrm{v}} R U(t) \, V^n(t) \,, \varphi \rangle \quad \text{on } [t_{n-1}, t_n],$$
$$\tag{6.29b}$$

for all $\varphi \in \mathbb{V}_n$. Note that the functions V^n, J^n are different from V respectively J: Whereas V and J are linear interpolations defined by the sequences $\{V_n\}_{n=1,\ldots,N}$ and $\{J_n\}_{n=1,\ldots,N}$, V^n and J^n are the (exact) solutions of above ODEs (6.29) on the time interval I_n.

To discretize (6.29) with respect to time, we employ the implicit Euler scheme, which reads: Find $V_n, J_n \in \mathbb{V}_n$ such that

$$\frac{1}{\tau_n} \langle J_n - J_{n-1} \,, \varphi \rangle + \langle |Z^{\mathrm{w}}| \, C^{\mathrm{m}} R C^{\mathrm{H}} U_n \, V_n \,, \varphi \rangle = 0, \qquad (6.30a)$$

$$\frac{1}{\tau_n} \langle V_n - V_{n-1} \,, \varphi \rangle + \langle C^{\mathrm{H}} m^{\mathrm{v}} R U_n \, V_n \,, \varphi \rangle = 0 \qquad (6.30b)$$

for all $\varphi \in \mathbb{V}_n$. As V_{n-1}, V_n and J_{n-1}, J_n define the linear interpolations of V respectively J on $[t_{n-1}, t_n]$, we see that V and J are actually time discrete approximations for V^n and J^n. To summarize this discretization of the ODEs and the discretization of the

PDE, we recall that the latter was given in (6.26) as

$$\left\langle (J_n |Z^{\mathrm{a}}| + |Z^{\mathrm{w}}|) \frac{U_n - U_{n-1}}{\tau_n} , \varphi \right\rangle + \langle J_n D^{\mathrm{a}} P^{\mathrm{a}} \nabla U_n , \nabla \varphi \rangle$$

$$+ \left\langle \left(|Z^{\mathrm{w}}| m^{\mathrm{u}} R V_n + |Z^{\mathrm{a}}| \frac{J_n - J_{n-1}}{\tau_n} \right) U_n , \varphi \right\rangle$$

$$+ C^{\mathrm{ext}} \langle U_n - u^{\mathrm{ext}} , \varphi \rangle_{\partial\Omega} = 0$$
$$(6.30c)$$

Equations (6.30b – 6.30c) are equations in finite dimensional spaces and we may thus solve them numerically. To that end, we observe that instead of considering equations (6.30) for *all* $\varphi \in \mathbb{V}_n$, we may equivalently resort to a basis $\{\varphi_1, \ldots, \varphi_{M_n}\} \subset \mathbb{V}_n$ of $\mathbb{V}_n$, $\dim \mathbb{V}_n = M_n$. Hence, (6.30) states a total of $3M_n$ equations, which are coupled non-linearly. Conveniently, we collect those equations by defining the non-linear functional

$$F_n : \mathbb{V}_n \times \mathbb{V}_n \times \mathbb{V}_n \to \mathbb{R}^{3M_n}.$$

Thereby, the components of $F_n(U_n, V_n, J_n) \in \mathbb{R}^{3M_n}$ are defined via the left hand sides of (6.30a), (6.30b), and (6.30c). More precisely, for arbitrary $\tilde{U}, \tilde{V}, \tilde{J} \in \mathbb{V}_n$, we define the components of $F_n(\tilde{U}, \tilde{V}, \tilde{J})$ as

$$(F_n(\tilde{U}, \tilde{V}, \tilde{J}))_i := \frac{1}{\tau_n} \langle \tilde{J} - J_{n-1} , \varphi_i \rangle + \langle |Z^{\mathrm{w}}| C^{\mathrm{m}} R C^{\mathrm{H}} \tilde{U} \, \tilde{V} , \varphi_i \rangle,$$

$$(F_n(\tilde{U}, \tilde{V}, \tilde{J}))_{M_n+i} := \frac{1}{\tau_n} \langle \tilde{V} - V_{n-1} , \varphi_i \rangle + \langle C^{\mathrm{H}} m^{\mathrm{v}} R \tilde{U} \, \tilde{V} , \varphi_i \rangle,$$

$$(F_n(\tilde{U}, \tilde{V}, \tilde{J}))_{2M_n+i} := \left\langle (\tilde{J} |Z^{\mathrm{a}}| + |Z^{\mathrm{w}}|) \frac{\tilde{U} - U_{n-1}}{\tau_n} , \varphi_i \right\rangle$$

$$+ \langle \tilde{J} D^{\mathrm{a}} P^{\mathrm{a}} \nabla \tilde{U} , \nabla \varphi_i \rangle$$

$$+ \left\langle \left(|Z^{\mathrm{w}}| m^{\mathrm{u}} R \tilde{V} + |Z^{\mathrm{a}}| \frac{\tilde{J} - J_{n-1}}{\tau_n} \right) \tilde{U} , \varphi_i \right\rangle$$

$$+ C^{\mathrm{ext}} \langle \tilde{U} - u^{\mathrm{ext}} , \varphi_i \rangle_{\partial\Omega},$$

for $i = 1, \ldots, M_n$. With this definition, equations (6.30a), (6.30b), and (6.30c) can equivalently be formulated as

$$F_n(U_n, V_n, J_n) = 0 \qquad (6.31)$$

and we employ Newton's method to solve (6.31) in every time step n.

Remark 6.3.1 (Convergence of Newton's method)
From the definition of F_n we see that F_n is continuously differentiable. Moreover, the Jacobian DF_n is Lipschitz continuous and assuming that $DF_n(U_n, V_n, J_n)$ is non-singular, Newton's method iteratively approximates the solution (U_n, V_n, J_n) of (6.31), provided an appropriate initial guess is chosen. In the computations of Section 6.5, Newton's method produced an adequate approximation of the solution of (6.31) in mostly two iterations.

6.3.4 Decoupled discretization

In this section, we present an alternate approach for computing $\{U_n\}_{n=1,\dots,N}$, $\{V_n\}_{n=1,\dots,N}$, and $\{J_n\}_{n=1,\dots,N}$, which is based on decoupling the PDE (6.26) and the ODEs of (6.27). To that end, we recall that the two ODEs (6.27a) and (6.27b) are particularly coupled to the PDE (6.26) as they involve u. We eliminate this coupling by considering the following modified versions of (6.27a) and (6.27b):

$$\frac{d}{dt} J^{\mathrm{a}}(t) = -|Z^{\mathrm{w}}| \, C^{\mathrm{m}} R C^{\mathrm{H}} u(t_{n-1}) \, v^{\mathrm{w}}(t) \qquad \text{on } [t_{n-1}, t_n],$$

$$\frac{d}{dt} v^{\mathrm{w}}(t) = -C^{\mathrm{H}} m^{\mathrm{v}} R u(t_{n-1}) \, v^{\mathrm{w}}(t) \qquad \text{on } [t_{n-1}, t_n],$$

where we substituted $u(t)$ by $u(t_{n-1})$. Note again, that in this formulation, on each time interval I_n, $n \geq 2$, the initial values $J^{\mathrm{a}}(t_{n-1})$ respectively $v^{\mathrm{w}}(t_{n-1})$ are given implicitly by the solution on the previous time interval. Only the values for $J^{\mathrm{a}}(0)$ and $v^{\mathrm{w}}(0)$ must be given explicitly. Resorting to a subspace $\mathbb{V}_n \subset H^1(\Omega)$ in each time step n, we consider the related problems: For $t \in [t_{n-1}, t_n]$, find

a) $J^n(t) \in \mathbb{V}_n$ such that

$$\frac{d}{dt}J^n(t) = -|Z^{\mathrm{w}}| \, C^{\mathrm{m}} R C^{\mathrm{H}} \mathcal{P}_n U_{n-1} V(t) \quad \text{on } [t_{n-1}, t_n],$$
$$(6.33a)$$
$$J^n(t_{n-1}) = \mathcal{P}_n J^{n-1}(t_{n-1});$$

b) $V^n(t) \in \mathbb{V}_n$ such that

$$\frac{d}{dt}V^n(t) = -C^{\mathrm{H}} m^{\mathrm{v}} R \mathcal{P}_n U_{n-1} V^n(t) \quad \text{on } [t_{n-1}, t_n],$$
$$(6.33b)$$
$$V^n(t_{n-1}) = \mathcal{P}_n V^{n-1}(t_{n-1}).$$

In this setting, we are no longer able to directly use $V^{n-1}(t_{n-1})$ — which is produced in time step $n-1$ — as an initial value for time step n since $V^{n-1}(t_{n-1}) \in \mathbb{V}_{n-1}$ and in general, the spaces $\mathbb{V}_{n-1} \not\subset \mathbb{V}_n$ are not nested. Instead, we employ a "representation" $\mathcal{P}_n V^{n-1}(t_{n-1}) \in \mathbb{V}_n$ as initial value. For the same reason, we also use a representation $\mathcal{P}_n U_{n-1} \in \mathbb{V}_n$ of $U_{n-1} \in \mathbb{V}_{n-1}$. The same considerations apply to the equation for $J^n(t)$.

In the implementation underlying the computational results presented in Section 6.5, we use the Lagrange interpolation operator $\mathcal{I}_{\mathbb{V}_n} : \mathbb{V}_{n-1} \to \mathbb{V}_n$ for representing U_{n-1}, $V^{n-1}(t_{n-1})$, and $J^{n-1}(t_{n-1})$ in $\mathbb{V}_n$, i. e., we set

$$\mathcal{P}_n U_{n-1} := \mathcal{I}_{\mathbb{V}_n} U_{n-1},$$
$$\mathcal{P}_n V^{n-1}(t_{n-1}) := \mathcal{I}_{\mathbb{V}_n} V^{n-1}(t_{n-1}),$$
$$\mathcal{P}_n J^{n-1}(t_{n-1}) := \mathcal{I}_{\mathbb{V}_n} J^{n-1}(t_{n-1}).$$

For solving the ODEs (6.33), we use the black box ODE solver RADAU5 by Hairer and Wanner, cf. [23].

Applying RADAU5 to (6.33b) first produces a solution $V^n(t)$ on the time interval I_n and we define the desired V_n as $V_n := V^n(t_n)$. With V_n known, the linear interpolation V is determined on the current time interval I_n and we may substitute this into (6.33a). Again, we use RADAU5 to solve (6.33a) in a second step. This produces the solution $J^n(t)$ on the time interval I_n and we define

$J_n := J^n(t_n)$. We then substitute the newly computed $V_n \in \mathbb{V}_n$ and $J_n \in \mathbb{V}_n$ into the PDE (6.26) and solve it, giving rise to the remaining desired $U_n \in \mathbb{V}_n$.

6.4 Implementation

In this section, we focus on the implementation of the two discretizations presented in Sections 6.3.3 and 6.3.4. First, we note that both presented discretizations involve a partition of the time interval as well as dedicated finite dimensional spaces $\mathbb{V}_n$ in each time step. The difference between the two discretizations is exclusively given in the equations which are to be solved in each time step.

Particularly, in the decoupled discretization, the discrete formulation (6.26) of the PDE considers V_n and J_n as given. In this setting, equation (6.26) strongly resembles the discretization (EG) of the linear parabolic problem (2.1) considered in Chapters 2 and 3. Moreover, considering v^w and J^a in the PDE (6.21b) as given, this equation is similar to the linear parabolic equation (2.1). The differences are only due to the fact that v^w and J^a are not constant but rather space and time dependent functions, as well as the boundary term reflecting the Robin boundary condition. Putting aside these rather minor differences, we are motivated to apply algorithm ASTFEM for solving problem (6.21). However, we emphasize that the quantities J^a and v^w are *not* known, but instead are the solutions of the two ODEs (6.21a) and (6.21c), respectively. In particular, these ODEs are coupled to the PDE (6.21b) since they both involve u. This particularly implies that the error control developed for the discretization of the linear parabolic equation is *not* valid here.

Nevertheless, as the process of concrete carbonation usually involves narrow reaction zones moving slowly through the material, we are motivated to use the *adaptive* algorithm ASTFEM for solving the carbonation problem numerically; see also [56] where adaptive simulations based on a simple reaction–diffusion model are presented. More precisely, the need for adaptivity is evoked by the fact that the sharp reaction front exhibits steep gradients of the involved concentrations while in other parts of the material, the concentrations vary only slightly or not at all. In order to capture such reaction fronts numerically, a high resolution — i. e., a fine mesh — is needed in the area of this front. This, of course, can be provided by employing a globally fine mesh which, however, is highly unwanted as it induces slow computations with high memory consumption. Moreover, a globally fine mesh is not needed since slowly varying concentrations can easily be approximated on a relatively coarse mesh. Thus, a method using a mesh which is (locally) fine near the reaction front and relatively coarse in other areas would make a good balance between accuracy and numerical cost.

For these reasons, moving fronts are actually the textbook example for problems which highly profit from employing adaptive methods. With this motivation, we slightly modify ASTFEM in the following in order to be able to solve the carbonation problem (6.21). In particular, we use linear finite elements for the spatial discretization but use a different module SOLVE and also redefine the error indicators.

Error indicators

We emphasize, that these redefined error indicators do not provide control over the discretization error, however, we may use them to drive the adaptive process heuristically. Moreover, we point out that we exclusively employ error indicators related to $U(t)$, i. e., the adaptive decisions heuristically aim at producing meshes $\mathcal{G}_n$ and time step sizes τ_n such that U is a good approximation of u. The redefined error indicators related to the coarsening and time error are

$$\mathcal{E}_{c\tau}^2(U_n, U_{n-1}) := D^{\mathrm{a}} P^{\mathrm{a}} \left\| J_n \nabla(U_n - U_{n-1}) \right\|^2,$$

$$\mathcal{E}_{\tau}^2(U_n, U_{n-1}, \mathcal{G}_n) := D^{\mathrm{a}} P^{\mathrm{a}} \left\| J_n \nabla(U_n - \Pi_n U_{n-1}) \right\|^2,$$

$$\mathcal{E}_c^2(U_{n-1}, \mathcal{G}_n) := D^{\mathrm{a}} P^{\mathrm{a}} \left\| J_n \nabla(U_{n-1} - \Pi_n U_{n-1}) \right\|^2,$$

compare also the original definitions for the linear parabolic problem in (4.2) on page 108. As a space error indicator, we use the standard residual error estimator related to the elliptic problem (6.26), where we regard U_n as the unknown and consider all other quantities as given. For the reader's convenience, we re-state (6.26), which reads:

$$\left\langle (J_n \, |Z^\mathrm{a}| + |Z^\mathrm{w}|) \, \frac{U_n - U_{n-1}}{\tau_n} \, , \, \varphi \right\rangle + \left\langle J_n D^\mathrm{a} P^\mathrm{a} \nabla U_n \, , \, \nabla \varphi \right\rangle$$

$$+ \left\langle \left(|Z^\mathrm{w}| \, m^\mathrm{u} R V_n + |Z^\mathrm{a}| \, \frac{J_n - J_{n-1}}{\tau_n} \right) U_n \, , \, \varphi \right\rangle$$

$$+ C^\mathrm{ext} \langle U_n - u^\mathrm{ext} \, , \, \varphi \rangle_{\partial\Omega} = 0$$

As indicated, we particularly regard J_n and J_{n-1} as well as V_n and U_{n-1} as given, however, we emphasize that they are functions. Since we are using *linear* finite elements, we particularly have for the second order term of the element residual

$$\mathrm{div}(D^\mathrm{a} P^\mathrm{a} J_n \nabla U_n) = D^\mathrm{a} P^\mathrm{a} \nabla J_n \cdot \nabla U_n + D^\mathrm{a} P^\mathrm{a} J_n \underbrace{\mathrm{div} \nabla U_n}_{=0}$$

$$= D^\mathrm{a} P^\mathrm{a} \nabla J_n \cdot \nabla U_n$$

and thus, the element residual of the standard residual error estimator for the stated elliptic problem is given by

$$R := - D^\mathrm{a} P^\mathrm{a} \nabla J_n \cdot \nabla U_n + \left(|Z^\mathrm{w}| \, m^\mathrm{u} R V_n + |Z^\mathrm{a}| \, \frac{J_n - J_{n-1}}{\tau_n} \right) U_n$$

$$+ (J_n \, |Z^\mathrm{a}| + |Z^\mathrm{w}|) \, \frac{U_n - U_{n-1}}{\tau_n}.$$

With this definition, we redefine the space error indicator as

$$\mathcal{E}_\mathcal{G}^2(U_n, U_{n-1}, \tau_n, f_n, \mathcal{G}_n, E) = h_E^2 \, \|R\|_{L^2(E)}^2$$

$$+ h_E \, \|[\![D^\mathrm{a} P^\mathrm{a} J_n \nabla U_n]\!]\|_{L^2(\partial E \cap \Omega)}^2$$

$$+ h_E \, \|J_n D^\mathrm{a} P^\mathrm{a} \nabla U_n \cdot \eta - C^\mathrm{ext}(U_n - u^\mathrm{ext})\|_{L^2(\partial\Omega)}^2.$$

Note that the redefined error indicators $\mathcal{E}_{c\tau}$, $\mathcal{E}_\tau$, and $\mathcal{E}_c$ additionally depend on the discrete function J_n. However, we do not state this dependency explicitly in the list of arguments for consistency with the notation of ASTFEM, but rather use the current value of J_n implicitly. The same applies to the error indicator $\mathcal{E}_\mathcal{G}$, where we omit stating the additional dependency on V_n and J_n explicitly.

Modules for solving

In order to apply ASTFEM for implementing the decoupled discretization, we also have to employ a modified module SOLVE. For the decoupled discretization, the new module is the following:

Algorithm 10 Module SOLVE for decoupled discretization

Employ RADAU5 to solve (6.33b) for $V^n(t)$

Set $V_n := V^n(t_n)$

Employ RADAU5 to solve (6.33a) for $J^n(t)$

Set $J_n := J^n(t_n)$

Solve (6.26) employing the newly computed V_n and J_n

For comparison, we are also motivated to solve the coupled discretization using ASTFEM and employing the error indicators defined above. Again, we emphasize that those error indicators provide no error control and are purely used as a heuristic for adapting time step sizes and meshes. This is particularly clear, as in the coupled discretization, no PDE is solved directly, but rather the nonlinear equation (6.31) — which represents the coupled system of the ODEs (6.30a), (6.30b) and the PDE (6.30c) — is solved. Accordingly, the module SOLVE simply reads:

Algorithm 11 Module SOLVE for coupled discretization

Employ Newton's method to solve (6.31)

Increasing the time step size

Another relevant aspect is the fact, that (6.26) does not involve a time dependent right hand side. Hence, the module CONSISTENCY in ASTFEM is not needed as no consistency error occurs. On the other hand, we realize that CONSISTENCY is the only module in ASTFEM which is able to enlarge the time step size. As this feature is very important also in the considered carbonation problem, we include the possibility for enlarging the time step size in the end of each time step, compare also the time step size enlargement in lines 27 through 29 of algorithm CLASSIC in Section 5.2. More precisely, if the time error indicator $\mathcal{E}_\tau$ was computed in the current time step, we check if $\mathcal{E}_\tau < \sigma \mathtt{TOL_S}$ is satisfied and if applicable, we enlarge the initial time step size for the next time step by a factor $\delta_2 > 1$. If $\mathcal{E}_\tau$ was not computed in the current time step — which is the case if the sum of space and coarsen-time error indicator satisfy the tolerance or if the space error indicator dominates — we also (unconditionally) enlarge the time step size. More precisely, we include the following lines between lines 14 and 15 of ASTFEM:

if $\mathcal{E}_\tau^2(U_n, U_{n-1}, \mathcal{G}_n)$ was *not* computed during this time step **then**

$$\tau_n = \delta_2 \tau_n$$

else if $\mathcal{E}_\tau^2(U_n, U_{n-1}, \mathcal{G}_n) < \sigma \mathtt{TOL_S}$ **then**

$$\tau_n = \delta_2 \tau_n$$

end if

Remark 6.4.1 (Enlarging vs. not enlarging τ in case of unknown $\mathcal{E}_\tau$)

Note that we might as well decide *not* to enlarge the time step size in case $\mathcal{E}_\tau$ was not computed in the current time step. As mentioned, $\mathcal{E}_\tau$ is not computed if either the sum of space and coarsen-time indicators sufficiently small or if the space indicator dominates. Especially in case of a dominating space error, it seems appropriate to vote in favor of an enlargement. Moreover, numerical experiments clearly show the advantages of enlarging the time step size in the considered case. The reason for this lies in the fact that $\mathcal{E}_\tau$ is only computed if the coarsen-time indicator dominates the space indicator. Since we do not employ a too harsh coarsening, this indicates quite a big time indicator $\mathcal{E}_\tau$, which in turn prevents τ from being enlarged. All together, deciding *not* to enlarge τ if $\mathcal{E}_\tau$ was not computed, leads to τ staying constant for a very long time. As the time step size is required to be very small in the beginning of the carbonation process, this implies a very slow time stepping. On the other hand, deciding to enlarge τ in the questionable case nicely enlarges the time step size and produces good results, see Section 6.5.

Tolerance distribution

Even though we do not employ the error indicators for error control, we recall from Section 4.3 that the error control for the linear parabolic problem involves the sum $\sum_{n=1}^{N} \tau_n \text{TOL}_S$, where TOL_S is the tolerance allowed in each time step. Given a total tolerance TOL and demanding

$$\sum_{n=1}^{N} \tau_n \text{TOL}_S \leq \text{TOL},$$

an obvious choice is to employ the same tolerance $\text{TOL}_S := \text{TOL}/T$ for each time step. However, this is not necessary and we may as well use individual tolerances TOL_S^n in each time step n as long as

$$\sum_{n=1}^{N} \tau_n \text{TOL}_S^n \leq \text{TOL}$$

is satisfied. Particularly, we define the tolerance for the n-th time step as

$$\text{TOL}_S^n := g_\alpha(t_n)\text{TOL} \qquad \text{with} \qquad g_\alpha(t) := \frac{1-\alpha}{T^{1-\alpha}}\, t^{-\alpha}, \quad \alpha \in (0,1).$$

Since $g_\alpha(t)$ is monotonically decreasing and the factor $(1-\alpha)/T^{1-\alpha}$ is chosen such that $\int_0^T g_\alpha(t)dt = 1$, we have

$$\sum_{n=1}^{N} \tau_n \text{TOL}_S^n = \sum_{n=1}^{N} \tau_n g_\alpha(t_n)\text{TOL} = \text{TOL} \sum_{n=1}^{N} \tau_n g_\alpha(t_n)$$

$$\leq \text{TOL} \int_0^T g_\alpha(t)\, dt = \text{TOL}.$$

Hence, provided a valid error control, employing the individual tolerances TOL_S^n achieves the same overall tolerance TOL as equidistributing TOL by setting $\text{TOL}_S^n = \text{TOL}/T$ for all $n = 1, \ldots, N$. In this sense, both tolerance distributions are equal.

In Section 6.5, we present computational results for both an equidistribution of the tolerance and an alternate tolerance distribution using $\text{TOL}_S^n := g_\alpha(t_n)\text{TOL}$ in the n-th time step with parameter $\alpha = 0.75$. The motivation for this lies in the nature of the carbonation process: In the beginning of the process, u locally possesses very steep gradients and requires a very fine mesh. However, as the carbonation evolves, the shape of u smoothens significantly and u can be resolved on quite a coarse mesh.

than the very simple assumptions made here so that the density of precipitated $CaCO_3$ in carbonated concrete is actually considerably smaller than it normally is.

We want to take into account this larger change of volume. For the simulations, we take a value of 13.5%. Under the assumption that the pore-water volume remains constant, this corresponds to a reduction of pore-air volume by 21%.

6.5.2 Interpretation of the results

Particularly, we focus on the simulation using the *decoupled* discretization and the *alternate distribution* of the tolerance with $\text{TOL}_\text{S}^n = g_\alpha(t_n)\text{TOL}$ and $\alpha = 0.75$. Based on this, Figure 6.3 shows the concentration of CO_2 after one year, four years, and 16 years in the top row. The middle row depicts the concentration of $Ca(OH)_2$ at the same times and the bottom row shows the pore-volume factor $J(t)$, which ranges from 0.79 (corresponding to the desired reduction of the pore-air volume by 21%) to 1. Whereas all concentrations are plotted in a normalized way, i.e., u/u^ext and v^w/v_0^w, we use dimensional quantities with respect to space and time. The long edges of the "L–shape" are five centimeters, the short edges are 2.5 centimeters. Note that the white line in Figure 6.3a indicates the position of the cut used later for concentration profiles and carbonation depths.

A more detailed comparison of the CO_2-concentrations after one and 16 years is shown in Figure 6.1. We observe, that the steep gradient in the CO_2-concentration after one year is greatly smoothened after 16 years. Opposed to that, the concentration of $Ca(OH)_2$, which is shown in Figure 6.2, exhibits steep gradients throughout the time interval. This implies that the bulk of the carbonation reaction is concentrated on a narrow zone which advances into the concrete with time. This coincides with other simulation results; cf. [49, 43, 34].

The smoothening of the CO_2 – concentration and the much steeper gradient of the $Ca(OH)_2$-concentration can also be seen in Figure 6.4, which shows profiles of the CO_2 and $Ca(OH)_2$-concentrations at a cut after 16 years. The orthogonal cut is located at the center of the left boundary and is indicated in Figure 6.3a.

We next focus on the *carbonation depth*, which is an impor-

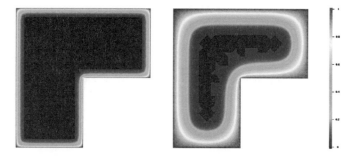

Figure 6.1: CO_2-concentration after one year (left) and 16 years (right).

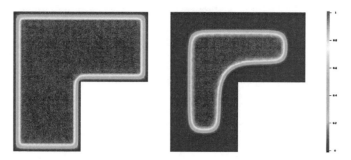

Figure 6.2: $Ca(OH)_2$-concentration after one year (left) and 16 years (right).

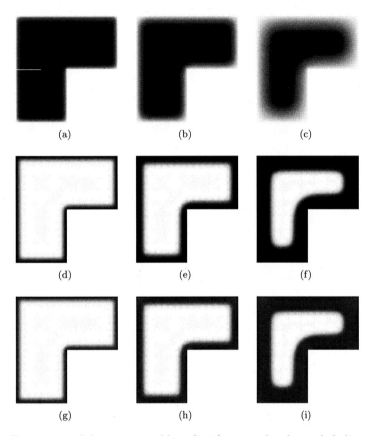

Figure 6.3: Solution to problem (6.21) using the decoupled discretization and alternately distributed tolerances. CO_2 (top) and $Ca(OH)_2$ (middle) concentrations as well as J^a (bottom) after 1 year (left), 4 years (middle), and 16 years (right). In sub-figures (a)–(f), black indicates value 0 and white indicates 1. In sub-figures (g)–(i), black indicates 0.79 and white indicates 1.

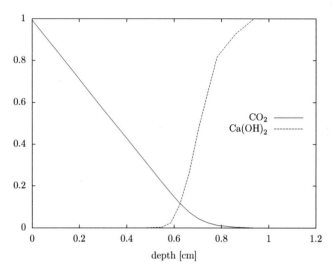

Figure 6.4: Concentration profile at cut from the decoupled discretization with alternate tolerance distribution.

tant quantity for durability issues. The carbonation depth is the depth the carbonation zone has penetrated into the concrete sample after a given amount of time. More precisely, we define the carbonation depth as the level set $v^w/v_0^w = 0.1$ following [61]. In Figure 6.5a, the predicted carbonation depth is plotted in comparison to experimental data of [67] (long-term exposure out of doors under roof). It can be seen that good approximation of experimental data is achieved. Furthermore, Figure 6.5a also compares the results with a simulation, where the evolution of the microstructure is neglected, i. e., the pore-volume factor J^a is constantly set to one. This results in an overestimation of the carbonation depth towards the end of the time interval, see also [39, 34]. Particularly, such an overestimation does not occur when considering the evolution of the microstructure.

For comparison, we also include the carbonation depth predicted by Peter and Böhm in [49] in Figure 6.5b and observe, that the results are indeed very similar. The carbonation depth in [49]

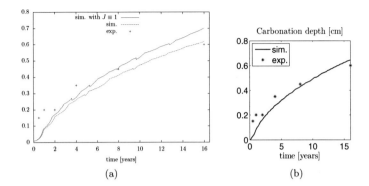

Figure 6.5: (a) Carbonation depth at cut for the presented model and an alternate model employing $J^a \equiv 1$. (b) For comparison, carbonation depth for the presented model from [49], courtesy of M. Peter.

is computed using a coupled discretization on a uniform mesh. For that reason we anticipate the comparison of the carbonation depth for decoupled and coupled discretization given in Figure 6.8 and notice that the results for the coupled discretization match the results of Peter and Böhm even better. For completeness, we mention that in order to provide sufficient resolution for the carbonation depth within an appropriate computation time, Peter and Böhm resorted to a small rectangular domain around the cut and employed appropriate Neumann boundary conditions. Opposed to that, the carbonation depth shown in Figure 6.5a is extracted from the solution over the whole domain. Appropriate resolution and computation time in this case is provided by decoupling the problem and most importantly, employing spatial adaptivity.

6.5.3 Comparison of the different discretizations

We now turn to comparing the different methods for solving the carbonation problem (6.21), which we motivated in Sections 6.3 and 6.4. In the process, we analyze both the coupled and the decoupled discretizations and employ in each case an equidistribution of the tolerance as well as the alternate distribution with

$\text{TOL}_S^n = g_\alpha(t_n)\text{TOL}$ with parameter $\alpha = 0.75$.

Equidistribution vs. alternate distribution of tolerance

The solutions of (6.21) employing an equidistribution of the tolerance are given in Figures 6.10 and 6.11. Opposed to that, Figures 6.3 and 6.9 show the results of using the alternately distributed tolerances. As indicated in the motivation of the alternate tolerance distribution (cf. page 223), we observe, that while the solutions after one and four years are resolved well also in case of equidistributed tolerance, the solutions after 16 years are not satisfactory. The reason for this are the very coarse meshes employed after 16 years, which only contain about 600 nodes, see Figures 6.14 and 6.15. Opposed to that, the solutions obtained by employing the alternate distribution of the tolerance, are resolved well for all times, see Figures 6.3 and 6.9. The corresponding meshes are depicted in Figures 6.12 and 6.13. Comparing them to the meshes used in the case of equidistributed tolerance, we observe that the most interesting zone, i. e., the reaction zone, is resolved much better in case of alternately distributed tolerance. This can also be seen by comparing Figures 6.6a and 6.7a, where concentration profiles are depicted. However, we emphasize, that whereas the reaction zone is characterized by steep gradients of the $Ca(OH)_2$-concentration, the gradients of the CO_2 concentrations in this zone are quite moderate. Since the employed adaptive procedure (heuristically) adapts the mesh with respect to the CO_2-concentration, we cannot expect a sufficient resolution of the reaction zone using equidistributed tolerances.

Coupled vs. decoupled discretization

Regarding the coupled and decoupled discretizations, we justify decoupling the system by observing that the results of the decoupled discretization (see Figures 6.3 and 6.10) coincide very well with the results of the coupled discretization (see Figures 6.9 and 6.11). This similarity can also be seen in the generated meshes, compare Figures 6.14 and 6.15, respectively Figures 6.12 and 6.13.

In Figures 6.6 and 6.7, we consider the concentration profiles for comparing the decoupled and coupled discretizations. The dis-

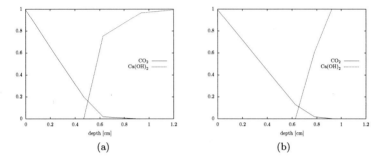

(a)   (b)

Figure 6.6: Concentration profile from decoupled discretization (a) and coupled discretization (b) with equidistribution of the tolerance.

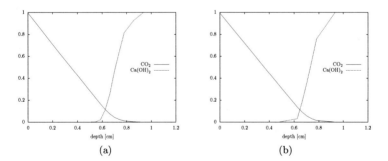

(a)   (b)

Figure 6.7: Concentration profile from decoupled discretization (a) and coupled discretization (b) with alternate tolerance distribution.

crepancy between the poorly resolved concentration profiles of Figures 6.6a and 6.6b are mostly due to differences in the meshes around the cutline. In case of the adequately resolved cutlines of Figure 6.7, we see that especially the concentration profiles for CO_2 coincide almost perfectly for the coupled and decoupled discretizations. This suggests that the employed decoupling indeed is legitimate. Moreover, this is also supported by the similar carbonation depths predicted using the coupled and decoupled discretization, see Figure 6.8.

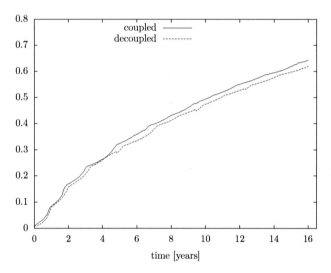

Figure 6.8: Carbonation depth at cut from coupled and decoupled discretization with alternate tolerance distribution.

Summarizing, we note that using a decoupled discretization of problem (6.21) is legitimate as it produces very similar results as the coupled discretization. Moreover, we may override the problem of insufficient resolution towards the end of the time interval — which is caused by adapting with respect to the CO_2-concentration instead of the $Ca(OH)_2$-concentration — by employing an alternate tolerance distribution. The profit of these methods in terms of computation time can be seen from Table 6.1, which particularly

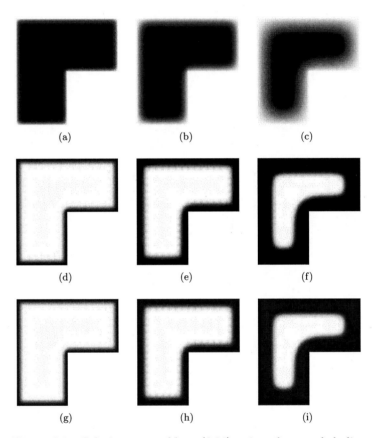

Figure 6.9: Solution to problem (6.21) using the coupled discretization with alternately distributed tolerances. CO_2 (top) and $Ca(OH)_2$ (middle) concentrations as well as J^a (bottom) after 1 year (left), 4 years (middle), and 16 years (right). In sub-figures (a)–(f), black indicates value 0 and white indicates 1. In sub-figures (g)–(i), black indicates 0.79 and white indicates 1.

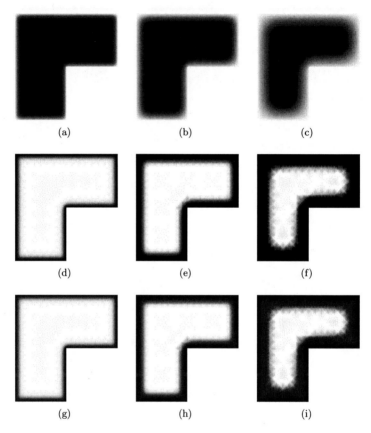

Figure 6.10: Solution to problem (6.21) using the decoupled discretization and equidistributed tolerance. CO_2 (top) and $Ca(OH)_2$ (middle) concentrations as well as J^a (bottom) after 1 year (left), 4 years (middle), and 16 years (right). In sub-figures (a)–(f), black indicates value 0 and white indicates value 1. In sub-figures (g)–(i), black indicates 0.79 and white indicates 1.

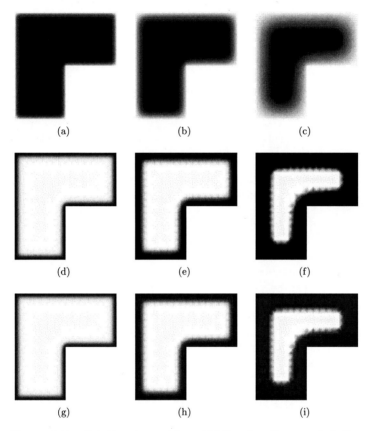

Figure 6.11: Solution to problem (6.21) using the coupled discretization and equidistributed tolerance. CO_2 (top) and $Ca(OH)_2$ (middle) concentrations as well as J^a (bottom) after 1 year (left), 4 years (middle), and 16 years (right). In sub-figures (a)–(f), black indicates value 0 and white indicates value 1. In sub-figures (g)–(i), black indicates 0.79 and white indicates 1.

	coupled	decoupled
equidistributed tolerance	2415.6	322.18
alternatively distributed tolerance	2113.2	202.66

Table 6.1: Computation times in seconds on an AMD Opteron 252 at 2.6GHz with 8GB RAM.

shows that using the decoupled discretization with alternatively distributed tolerance provides a speedup of nearly factor 12 compared to the coupled discretization with equidistributed tolerance.

Moreover, we point out that the meshes depicted in Figures 6.12, 6.13, 6.14, and 6.15 clearly show that the considered carbonation problem highly profits from employing an adaptive method. Particularly the meshes obtained when using the alternatively distributed tolerance (see Figures 6.12 and 6.13) show that in this case, the movement of the narrow reaction zone is captured well by producing meshes which are very fine in this area. Aiming at reaching results of similar quality using a non-adaptive approach results in a uniform mesh which is globally comparably fine as the adaptive meshes only are near the reaction front. We point out that this uniform mesh is then used throughout the time interval and contains about 50.000 nodes. Compared to the adaptively generated meshes, which only contain about 1.000 − 5.000 nodes, this tremendously increases both computational time and memory consumption.

6.6 Conclusions and outlook

We have presented an adaptive space time finite element method (ASTFEM) for the numerical solution of linear parabolic partial differential equations, which reaches any prescribed positive tolerance within a finite number of iterations. This method is based on a new approach of error control taking advantage of a uniform energy estimate. Moreover, we have introduced a flexible coarsening strategy and presented numerical examples employing this strategy in ASTFEM, which show that the introduced adaptive method performs well. Further, we applied a modified version of ASTFEM'

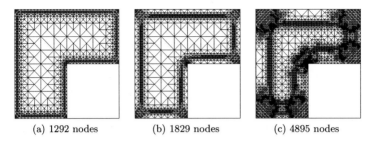

| (a) 1292 nodes | (b) 1829 nodes | (c) 4895 nodes |

Figure 6.12: Meshes corresponding to the decoupled discretization with alternately distributed tolerances depicted in Figure 6.3.

to the real life problem of concrete carbonation, where we considered and compared two different discretizations and presented numerical results.

Particularly where the simulations of the concrete carbonation are concerned, future work might include spatial adaptation with respect to a quantity taking into account the concentration of $Ca(OH)_2$, which exhibits much steeper gradients than the concentration of CO_2. This might lead to a spatial adaptation performing well also in case of a (temporally) equidistributed tolerance.

With respect to the theoretical parts of this thesis, future research might focus on the derivation of a terminating adaptive method without employing an a priori minimal time step size. This is particularly motivated by the fact that the minimal time step size is never reached during numerical tests using ASTFEM. This behavior suggests that improving and extending the analysis of ASTFEM might reveal a theoretical principle allowing for discarding the minimal time step size while still guaranteeing convergence into tolerance within finitely many iterations.

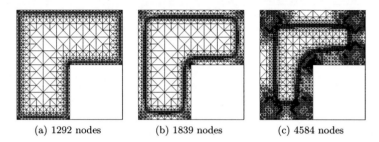

(a) 1292 nodes   (b) 1839 nodes   (c) 4584 nodes

Figure 6.13: Meshes corresponding to the coupled discretization with alternately distributed tolerances distribution depicted in Figure 6.9.

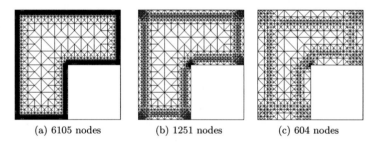

(a) 6105 nodes   (b) 1251 nodes   (c) 604 nodes

Figure 6.14: Meshes corresponding to the decoupled discretization with equidistributed tolerance depicted in Figure 6.10.

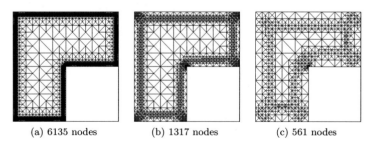

(a) 6135 nodes   (b) 1317 nodes   (c) 561 nodes

Figure 6.15: Meshes corresponding to the coupled discretization with equidistributed tolerance depicted in Figure 6.11.

Bibliography

[1] R. A. Adams and J. J. F. Fournier. *Sobolev spaces. 2nd ed.* Pure and Applied Mathematics 140. New York, NY: Academic Press. xiii, 305 p., 2003.

[2] M. Ainsworth and J. Oden. *A posteriori error estimation in finite element analysis.* Pure and Applied Mathematics. A Wiley-Interscience Series of Texts, Monographs, and Tracts. Chichester: Wiley. xx, 2000.

[3] H. W. Alt. *Linear functional analysis. An application oriented introduction. (Lineare Funktionalanalysis. Eine anwendungsorientierte Einführung.) 4., überarb. und erweiterte Aufl.* Berlin: Springer. xiv, 2002.

[4] D. N. Arnold, A. Mukherjee, and L. Pouly. Locally adapted tetrahedral meshes using bisection. *SIAM J. Sci. Comput.*, 22(2):431–448, 2000.

[5] B. Aulbach and T. Wanner. Integral manifolds for Carathéodory type differential equations in Banach spaces. Aulbach, B. (ed.) et al., Six lectures on dynamical systems. Papers from the tutorial workshop, Augsburg, Germany, June 1994. Singapore: World Scientific. 45-119, 1996.

[6] E. Bänsch. Local mesh refinement in 2 and 3 dimensions. *IMPACT Comput. Sci. Eng.*, 3(3):181–191, 1991.

[7] A. Bensoussan, J.-L. Lions, and G. Papanicolaou. *Asymptotic analysis for periodic structures.* North-Holland, 1978.

[8] T. A. Bier. *Karbonatisierung und Realkalisierung von Zementstein und Beton.* PhD thesis, University of Karlsruhe, 1988.

[9] D. Braess. *Finite Elemente.* Springer Verlag, 2007.

[10] J. H. Bramble, J. E. Pasciak, and O. Steinbach. On the stability of the L^2 projection in $H^1(\Omega)$. *Math. Comput.*, 71(237):147–156, 2002.

[11] S. C. Brenner and L. R. Scott. *The Mathematical Theory of Finite Element Methods.* Springer Verlag, 1994.

[12] D. Bunte. *Zum Karbonatisierungsbedingten Verlust der Dauerhaftigkeit von Aussenbauteilen aus Stahlbeton.* PhD thesis, Technical University of Braunschweig, 1994.

[13] T. Chaussadent. États de lieux et réflexions sur la carbonatation du beton armé. Technical Report LCPC OA29, Laboratoire Central de Ponts et Chaussées, Paris, 1999.

[14] Z. Chen and J. Feng. An adaptive finite element algorithm with reliable and efficient error control for linear parabolic problems. *Math. Comput.*, 73(247):1167–1193, 2004.

[15] P. G. Ciarlet. *The Finite Element Method for Elliptic Problems.* North-Holland Publishing Company, 1978.

[16] D. Cioranescu and P. Donato. *An introduction to homogenization.* Oxford University Press, 1999.

[17] P. Clément. Approximation by finite element functions using local regularization. *R.A.I.R.O. Analyse Numrique R-2, pp.77-84*, 1975.

[18] R. Dautray and J.-L. Lions. *Mathematical analysis and numerical methods for science and technology. Vol. 5: Evolution problems I.* Berlin: Springer. xiv, 2000.

[19] L. C. Evans. *Partial Differential Equations.* Graduate Studies in Mathematics. 19. Providence, AMS, 1998.

[20] D. Gilbarg and N. S. Trudinger. *Elliptic partial differential equations of second order. Reprint of the 1998 ed.* Classics in Mathematics. Berlin: Springer. xiii, 2001.

[21] P. Grisvard. *Elliptic problems in nonsmooth domains.* Monographs and Studies in Mathematics, 24. Pitman Advanced Publishing Program. Boston-London-Melbourne: Pitman Publishing Inc. XIV, 410 p. , 1985.

[22] W. Hackbusch. *Elliptic differential equations: theory and numerical treatment.* Springer Series in Computational Mathematics. 18. Berlin: Springer- Verlag. xiv, 1992.

[23] E. Hairer and G. Wanner. *Solving ordinary differential equations. II: Stiff and differential-algebraic problems. Reprint of the 1996 2nd revised ed.* Springer Series in Computational Mathematics 14. Berlin: Springer. xvi,, 2010.

[24] J. Jost. *Partial differential equations. 2nd ed.* Graduate Texts in Mathematics 214. New York, NY: Springer. xiii, 356 p. , 2007.

[25] P. Knabner and L. Angermann. *Numerical methods for elliptic and parabolic partial differential equations.* Texts in Applied Mathematics 44. New York, NY: Springer. xv, 2003.

[26] I. Kossaczký. A recursive approach to local mesh refinement in two and three dimensions. *J. Comput. Appl. Math.*, 55(3):275–288, 1994.

[27] C. Kreuzer, C. A. Möller, K. G. Siebert, and A. Schmidt. Design and convergence analysis for an adaptive discretization of the heat equation. *in preparation.*

[28] J. Kropp. Relations between transport characteristics and durability. In J. Kropp and H. K. Hilsdorf, editors, *Performance criteria for concrete durability*, number 12 in RILEM Report, pages 97–137. E & FN SPON, 1995.

[29] D. R. Lide, editor. *CRC Handbook of Chemistry and Physics.* CRC Press LLC, 82 edition, 2001.

[30] A. Liu and B. Joe. Quality local refinement of tetrahedral meshes based on bisection. *SIAM J. Sci. Comput.*, 16(6):1269–1291, 1995.

[31] A. S. M. Böhm, A. Muntean. On the motion of internal carbonation layers.

[32] J. M. Maubach. Local bisection refinement for n-simplicial grids generated by reflection. *SIAM J. Sci. Comput.*, 16(1):210–227, 1995.

[33] S. A. Meier, M. A. Peter, and M. Böhm. A two-scale modelling approach to reaction–diffusion processes in porous materials. *Comp. Mat. Sci.*, 39(1):29–34, 2007.

[34] S. A. Meier, M. A. Peter, A. Muntean, and M. Böhm. Modelling and simulation of concrete carbonation with internal layers. *Chem. Eng. Sci.*, 62 (2007), pages 1125–1137, 2007.

[35] C. A. Möller, M. A. Peter, and K. G. Siebert. Multi-scale modelling and adaptive finite element simulation of concrete carbonation. *in preparation.*

[36] P. Morin, K. G. Siebert, and A. Veeser. A basic convergence result for conforming adaptive finite elements. *Math. Models Methods Appl. 18 (2008)*, 18:707–737, 2008.

[37] A. Muntean. *A Moving-Boundary Problem: modeling, analysis and simulation of concrete carbonation.* PhD dissertation, University of Bremen. Cuvillier, 2006.

[38] A. Muntean. Error bounds on semi-discrete finite element approximations of a moving-boundary system arising in concrete corrosison. *International Journal of Numerical Analysis and Modeling, 5(3):363-372*, 2008.

[39] A. Muntean, S. A. Meier, M. A. Peter, M. Böhm, and J. Kropp. A note on the limitations of the use of accelerated concrete-carbonation tests for service-life predictions. Berichte aus der Technomathematik 05-04, ZeTeM, University of Bremen, 2005.

[40] R. H. Nochetto, K. G. Siebert, and A. Veeser. Theory of adaptive finite element methods: An introduction. In R. A. DeVore and A. Kunoth, editors, *Multiscale, Nonlinear and Adaptive Approximation*, pages 409–542. Springer, 2009.

[41] M. Panfilov. *Macroscale models of flow through highly heterogeneous porous media*. Kluwer Academic, 2000.

[42] L. Pankratov, A. Piatnitskii, and V. Rybalko. Homogenized model of reaction–diffusion in a porous medium. *C. R. Mécanique*, 331:253–258, 2003.

[43] V. G. Papadakis, C. G. Vayenas, and M. N. Fardis. A reaction engineering approach to the problem of concrete carbonation. *AIChE J.*, 35(10):1639–1650, 1989.

[44] M. A. Peter. Homogenisation in domains with evolving microstructure. *C. R. Mécanique*, 335(7):357–362, 2007.

[45] M. A. Peter. Homogenisation of a chemical degradation mechanism inducing an evolving microstructure. *C. R. Mécanique*, 335(11):679–684, 2007.

[46] M. A. Peter. Coupled reaction–diffusion processes inducing an evolution of the microstructure: analysis and homogenization. *Nonlin. Anal*, 70(2):806–821, 2009.

[47] M. A. Peter and M. Böhm. Scalings in homogenisation of reaction, diffusion and interfacial exchange in a two-phase medium. In M. Fila, A. Handlovicova, K. Mikula, M. Medved, P. Quittner, and D. Sevcovic, editors, *Proc. Equadiff-11*, pages 369–376, 2005.

[48] M. A. Peter and M. Böhm. Different choices of scaling in homogenization of diffusion and interfacial exchange in a porous medium. *Math. Meth. Appl. Sci.*, 31(11):1257–1282, 2008.

[49] M. A. Peter and M. Böhm. Multi-scale modelling of chemical degradation mechanisms in porous media with evolving microstructure. *SIAM Multisc. Mod. Sim.*, 7(4):1643–1668, 2009.

[50] M. A. Peter, A. Muntean, S. A. Meier, and M. Böhm. Competition of several carbonation reactions in concrete: a parametric study. *Cem. Concr. Res.*, 38(12):1385–1393, 2008.

[51] A. V. Saetta, B. A. Schrefler, and R. V. Vitaliani. The carbonation of concrete and the mechanism of moisture, heat and carbon dioxide flow through porous materials. *Cem. Concr. Res.*, 23(4):761–772, 1993.

[52] E. Samson, J. Marchand, and J. J. Beaudoin. Describing ion diffusion mechanisms in cement-based materials using the homogenization technique. *Cem. Concr. Res.*, 29(8):1341–1345, 1999.

[53] E. Sanchez-Palencia. *Non-homogeneous media and vibration theory*. Springer, 1980.

[54] A. Schmidt. A multi-mesh finite element method for 3D phase field simulations. Colli, Pierluigi (ed.) et al., Free boundary problems: theory and applications. Proceedings of a conference, Trento, Italy, June 2002. Basel: Birkhäuser. ISNM, Int. Ser. Numer. Math. 147, 293-301 (2003)., 2003.

[55] A. Schmidt. A multi-mesh finite element method for phase-field simulations. Emmerich, Heike (ed.) et al., Interface and transport dynamics. Computational modelling. International workshop on computational physics of transport and interfacial dynamics, Dresden, Germany, February 25 to March 8, 2002. Proceedings. Berlin: Springer. Lect. Notes Comput. Sci. Eng. 32, 208-217 (2003)., 2003.

[56] A. Schmidt, A. Muntean, and M. Böhm. Numerical experiments with self-adaptive finite element simulations in 2d for the carbonation of concrete. *Report 05-01, ZeTeM University of Bremen*, 2005.

[57] A. Schmidt and K. G. Siebert. *Design of adaptive finite element software. The finite element toolbox ALBERTA*. Lecture Notes in Computational Science and Engineering 42. Berlin: Springer. xii, 2005.

[58] L. Scott and S. Zhang. Finite element interpolation of nonsmooth functions satisfying boundary conditions. *Math. Comput.*, 54(190):483–493, 1990.

[59] K. G. Siebert. A convergence proof for adaptive finite elements without lower bound. *IMA Journal of Numerical Analysis*, 2010.

[60] K. Sisomphon. *Influence of pozzolanic material additions on the development of the alkalinity and the carbonation behaviours of composite cement pastes and concretes.* PhD thesis, Technical University Hamburg–Harburg, 2004.

[61] A. Steffens, D. Dinkler, and H. Ahrens. Modeling carbonation for corrosion risk prediction of concrete structures. *Cem. Concr. Res.*, 32(6):935–941, 2002.

[62] R. Stevenson. The completion of locally refined simplicial partitions created by bisection. *Math. Comput.*, 77(261):227–241, 2008.

[63] R. Verfürth. A posteriori error estimation and adaptive mesh-refinement techniques. *Journal of Computational and Applied Mathematics*, 50(1-3):67 – 83, 1994.

[64] R. Verfürth. *A review of a posteriori error estimation and adaptive mesh-refinement techniques.* Wiley-Teubner Series Advances in Numerical Mathematics. Chichester: John Wiley & Sons. Stuttgart: B. G. Teubner. vi, 1996.

[65] R. Verfürth. A posteriori error estimates for finite element discretizations of the heat equation. *Calcolo*, 40(3):195–212, 2003.

[66] D. Werner. *Functional analysis. (Funktionalanalysis.) 6th corrected ed.* Springer-Lehrbuch. Berlin: Springer. xiii, 2007.

[67] H.-J. Wierig. Longtime studies on the carbonation of concrete under normal outdoor exposure. *Proceedings of the RILEM, Hannover University, Hannover, Germany (1984)*, pages 239–249, 1984.

[68] J. Wloka. *Partielle Differentialgleichungen.* B. G. Teubner, Stuttgart, 1982.

[69] K. Yosida. *Functional analysis. 6th ed.* Grundlehren der mathematischen Wissenschaften, 123. Berlin-Heidelberg-New York: Springer-Verlag. XII, 1980.

Lebenslauf

Name   Christian A. Möller
                geboren 1983

Ausbildung

2003   Abitur, Allgäu-Gymnasium Kempten

2003 – 2005   Studium Mathematik mit Nebenfach Physik,
                Universität Augsburg

2005 – 2010   Studium im Elitestudiengang TopMath,
                Universität Augsburg

2007 – 2010   Promotion am Lehrstuhl für Angewandte
                Analysis mit Schwerpunkt Numerik der
                Universität Augsburg im Rahmen von TopMath